Anatomy of Flowering Plants

Understanding plant anatomy is not only fundamental to the study of plant systematics and palaeobotany, but is also an essential part of evolutionary biology, physiology, ecology, and the rapidly expanding science of developmental genetics. In the third edition of her successful textbook, Paula Rudall provides a comprehensive yet succinct introduction to the anatomy of flowering plants. Thoroughly revised and updated throughout, the book covers all aspects of comparative plant structure and development, arranged in a series of chapters on the stem, root, leaf, flower, seed and fruit. Internal structures are described using magnification aids from the simple hand-lens to the electron microscope. Numerous references to recent topical literature are included, and new illustrations reflect a wide range of flowering plant species. The phylogenetic context of plant names has also been updated as a result of improved understanding of the relationships among flowering plants. This clearly written text is ideal for students studying a wide range of courses in botany and plant science, and is also an excellent resource for professional and amateur horticulturists.

Paula Rudall is Head of Micromorphology (Plant Anatomy and Palynology) at the Royal Botanic Gardens, Kew. She has published more than 150 peer-reviewed papers, using comparative floral and pollen morphology, anatomy and embryology to explore evolution across seed plants.

Anatomy of Flowering Plants

An Introduction to Structure and Development

PAULA J. RUDALL

CAMBRIDGE
UNIVERSITY PRESS

CAMBRIDGE UNIVERSITY PRESS
Cambridge, New York, Melbourne, Madrid, Cape Town, Singapore,
São Paulo

Cambridge University Press
The Edinburgh Building, Cambridge CB2 2RU, UK

Published in the United States of America by Cambridge University Press,
New York

www.cambridge.org
Information on this title: www.cambridge.org/9780521692458

Third edition published 2007

Printed in the United Kingdom at the University Press, Cambridge

A catalogue record for this publication is available from the British Library

Library of Congress Cataloguing in Publication data

ISBN-13 978-0-521-69245-8 paperback
ISBN-10 0-521-69245-8 paperback

Contents

Preface

In the twenty-first century, plant anatomy remains highly relevant to systematics, paleobotany, and the relatively new science of developmental genetics, which interfaces disciplines and utilizes a combination of techniques to examine gene expression in growing tissues. Modern students need to consider information from an increasingly wide range of sources, most notably integrating morphological and molecular data. The third, thoroughly revised, edition of this book presents an introduction to plant anatomy for students of botany and related disciplines.

Although the simple optical lens has been used for centuries to examine plant structure, detailed studies of plant anatomy originated with the invention of the compound microscope in the seventeenth century. Nehemiah Grew (1641—1712) and Marcello Malpighi (1628—1694), physicians working independently in England and Italy respectively, were early pioneers of the microscopical examination of plant cells and tissues. Their prescient work formed the foundation that eventually led to the development of our understanding of cell structure and cell division[27]. Other early outstanding figures included Robert Brown (1773—1858), who discovered the nucleus, and the plant embryologist Wilhelm Hofmeister (1824—1877), who first described the alternation of generations in the life cycle of land plants. In the nineteenth and twentieth centuries plant anatomy became an important element of studies of both physiology and systematic biology, and an integral aspect of research in the developing field of anatomical paleobotany, led by such luminaries as Dukinfield Henry Scott (1854—1934). The physiologist Gottlieb Haberlandt (1854—1945) utilized anatomical observations in his ground-breaking work on photosynthetic carbon metabolism. One of the most notable plant anatomists of the twentieth century was Katherine Esau (1898—1997), recognized particularly for her work on the structure and development of phloem and her influential textbooks on plant anatomy[30]. Other important textbooks include works on paleobotany, morphology, anatomy and embryology[13,34,68,106].

The invention of the transmission electron microscope (TEM) in the mid twentieth century allowed greater magnification than any optical microscope, and hence revitalized studies in cell ultra-structure and pollen morphology[98]. The subsequent invention of the scanning electron microscope (SEM) provided greater image clarity and much greater depth of focus than light microscopes, and thus further increased accessibility of minute structures, including seeds, pollen grains and organ primordia[28,98]. More recent innovations, including fluorescence microscopy, differential interference contrast (DIC) microscopy and confocal imaging, have allowed enhanced visualization of tissue structure. Others, including nuclear magnetic resonance (NMR) imaging and high-resolution X-ray computed tomography (HRCT) facilitate enhanced visualization of three-dimensional objects.

Taxonomic Overview

In textbooks published before 1990, extant angiosperms were consistently subdivided into two major groups – dicotyledons (dicots) and monocotyledons (monocots), based partly on the number of cotyledons in the seedling. This dichotomy was long considered to represent a fundamental divergence at the base of the angiosperm evolutionary tree. Other features marked this distinction, including the absence of a vascular cambium and presence of parallel leaf venation in monocots. However, the expansion of molecular phylogenetics through the early 1990s indicated that some species that were formerly classified as primitive dicots do not belong to either category, though the monophyly of monocots was confirmed[2,3,103]. Thus, although the dicot/monocot distinction remains useful for generalized descriptions of angiosperm groups, current evidence suggests that it does not represent a wholly natural classification. It is now widely accepted that several relatively species-poor angiosperm lineages (here termed early-divergent angiosperms or magnoliids) evolved before the divergence of the two major lineages that led to the monocots and the remaining dicots (now termed eudicots, or sometimes tricolpates).

Early-divergent angiosperms (including magnoliids) are a small but highly diverse assemblage of taxonomically isolated lineages that probably represent the surviving extant members of their respective clades, accounting for only about 1% of extant species. They possess some morphological features in common with both monocots and eudicots, and include the New Caledonian shrub *Amborella*, the water lilies (Nymphaeaceae), woody families such as Magnoliaceae and Lauraceae, and herbaceous or climbing families such as Piperaceae and Aristolochiaceae. Monocots account for approximately a quarter of all flowering plants species. They dominate significant parts of world ecosystems, and are of immense economic importance, including the staple grass food crops (wheat, barley, rice and maize) and other important food plants such as onions, palms, yams, bananas and gingers. Eudicots represent about 75% of extant angiosperm species, and encompass a wide range of morphological diversity, especially in the two largest subclades, Rosidae (rosid eudicots) and Asteridae (asterid eudicots).

1

Organs, cells and tissues

1.1 Organs

Plants consist of several organs, which in their turn are composed of tissues. Broadly, vegetative organs support plant growth, and reproductive organs enable sexual reproduction. The three main types of vegetative organ are the root, stem and leaf. Roots typically occur underground, and extract moisture and nutrients from the soil, though there are many examples of plants with aerial roots. The stem and leaves together comprise the shoot (Fig. 1.1). Stems occur both above and below ground. Some stems are modified into underground perennating or storage organs such as corms or rhizomes. Leaves typically occur above ground level, though some underground stems possess reduced scale leaves, and underground bulbs possess swollen leaves or leaf bases.

Primary organs and tissues develop initially from the shoot and root apical meristems and from cell divisions in meristems closely adjacent to them, such as the primary thickening meristem. Secondary tissues such as secondary xylem (wood) develop from lateral meristems such as the vascular cambium. Organs such as adventitious roots develop from differentiated cells that have retained meristematic capacity. At the onset of flowering, the shoot apical meristem undergoes structural modification from a vegetative to a reproductive apex and subsequently produces flowers (chapter 5). Flowers are borne on an inflorescence, either in groups or as solitary structures. A group of inflorescences borne on a single plant is termed a synflorescence[121] (Fig. 1.2).

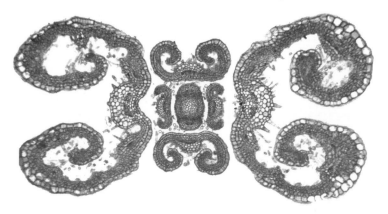

Figure 1.1 *Hyptis ditassoides* (Lamiaceae), transverse section of vegetative bud near apex, showing three successive pairs of leaf primordia surrounding central stem. Scale = 100 μm.

1.2 Cells

Plant cells typically have a cell wall containing a living protoplast (Fig. 1.3). The layer that contacts the walls of adjacent cells is termed the middle lamella. Following cessation of growth, many cells develop a secondary cell wall which is deposited on the inside surface of the primary wall. Both primary and secondary walls consist of cellulose microfibrils embedded in a matrix and oriented in different directions. Secondary cell walls consist mostly of cellulose, but primary walls commonly contain a high proportion of hemicelluloses in the gel-like matrix, affording a greater degree of plasticity to the wall of the growing cell. The secondary wall can also contain deposits of lignin (in sclerenchymatous cells) or suberin (in many periderm cells), and often appears lamellated.

Thin areas of the primary wall, which usually correspond with thin areas of the walls of neighbouring cells, are primary pit fields, and usually have protoplasmic strands (plasmodesmata) passing through them, connecting the protoplasts of neighbouring cells[36]. The connected living protoplasts are collectively termed the symplast. Primary pit fields often remain as thin areas of the wall even after a secondary wall has been deposited, and are then termed

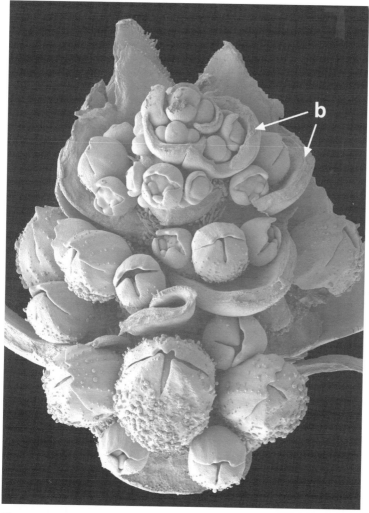

Figure 1.2 *Salvia involucrata* (Lamiaceae), dissected developing synflorescence showing flower clusters, each consisting of three flowers enclosed within a bract; younger stages towards apex. b = bract. Scale = 500 μm.

pits, or pit-pairs if there are two pits connecting adjacent cells. Pits may be simple, as in most parenchyma cells, or bordered, as in tracheary elements. In simple pits the pit cavity is of more or less uniform width, whereas in bordered pits the secondary wall

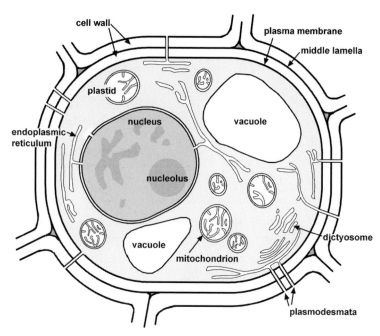

Figure 1.3 Diagram of a generalised plant cell illustrating details of protoplasmic contents.

arches over the pit cavity so that the opening to the cavity is relatively narrow. Through a light microscope the outer rim of the primary pit field appears as a border around the pit opening.

The cell protoplast is contained within a plasma membrane. It consists of cytoplasm that encloses bodies such as the nucleus, plastids and mitochondria, and also non-protoplasmic contents such as oil, starch or crystals. The nucleus, which is bounded by a nuclear membrane, often contains one or more recognizable bodies (nucleoli) together with the chromatin in the nuclear sap. During cell division the chromatin becomes organized into chromosomes. Most cells possess a single nucleus, but examples of multinucleate cells (coenocytes) include the non-articulated laticifers found in many plant families (chapter 1.4). Such cells elongate and penetrate established tissues by intrusive tip growth, in which the cell apices secrete enzymes that dissolve the middle lamellae of neighbouring cells; bifurcation occurs when they encounter an obstacle[36].

Mitochondria and plastids are surrounded by double membranes. Plastids are larger than mitochondria, and are classified into different types depending on their specialized role. For example, chloroplasts are plastids that contain chlorophyll within a system of lamellae that are stacked to form grana; this is the site of photosynthesis. Chloroplasts occur in all green cells, but are most abundant in the leaf mesophyll, which is the primary photosynthetic tissue (chapter 4.5). Membranes occur widely throughout the cytoplasm, sometimes bounding a series of cavities. For example, the endoplasmic reticulum is a continuous membrane-bound system of flattened sacs and tubules, sometimes coated with ribosomal particles. Dictyosomes are systems of sacs associated with secretory activity. Vacuoles are cavities in the cytoplasm; they are usually colourless and contain a watery fluid. Their size and shape varies in different cell types, and also changes during the life of a cell.

1.3 Cell Inclusions

Many cells possess non-protoplasmic contents such as oils, mucilage (slime), tannins, starch granules, calcium oxalate crystals and silica bodies. Both oil and mucilage are produced in secretory idioblasts which are often larger than adjacent parenchymatous cells. Tannins are phenol derivatives which are common in plant cells; they are amorphous, and appear yellow, red or brown in colour in cells of sectioned material. Cystoliths are cellulose bodies encrusted with calcium carbonate that occur in epidermal cells in some species (Fig. 4.4); the body is attached to the cell wall by a silicified stalk[36].

Starch is especially common in storage tissues such as endosperm or in parenchyma adjacent to a nectary. Starch granules

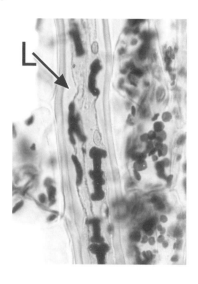

Figure 1.4 *Monadenium ellenbeckii* (Euphorbiaceae). Elongated I-shaped starch grains in laticifer (L); ovoid starch grains present in adjacent parenchyma cells. Scale = 20 μm.

are formed in plastids (amyloplasts). They often appear layered due to the successive deposition of concentric rings, and may possess characteristic shapes. For example, in species of *Euphorbia*, starch grains in laticifers are elongated and sometimes rod-shaped or bone-shaped compared with the more rounded starch grains of neighbouring parenchyma cells (Fig. 1.4)[70].

Calcium oxalate crystals (Figs 1.5, 1.13) are borne in crystal idioblasts that can occur in almost every part of the plant, including both vegetative and reproductive organs[82]. They are often present near veins, possibly due to transport of calcium through the xylem, and are sometimes associated with air space formation; some aquatic plants possess calcium oxalate crystals projecting into air spaces. Crystals form within vacuoles of actively growing cells and are usually associated with membrane chambers, lamellae, mucilage and fibrillar material. Crystal sand is relatively amorphous and represents fragmented non-nucleated crystalline particles. Druses (cluster crystals) are aggregated crystalline structures that have precipitated around a nucleation site. Raphides are bundles of needle-like crystals that are borne in the same cell; they occur commonly in monocots. In the monocot family Araceae,

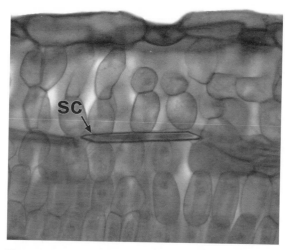

Figure 1.5 *Crocus cancellatus* (Iridaceae), longitudinal section of leaf showing crystal idioblast containing styloid crystal (sc). Scale = 50 μm.

raphides are characteristically grooved and sometimes barbed. Styloid crystals are typically solitary, larger and needle-like or rhomboidal; they are highly characteristic of some families, such as Iridaceae[91].

Opaline silica bodies are also a characteristic feature of some plant groups[83]. They occur in all plant parts, often associated with sclerenchyma, though they are rare in roots. In many dicot species they occur in the ray or axial parenchyma cells in secondary xylem. Some families, such as grasses (Poaceae), sedges (Cyperaceae), orchids (Orchidaceae) and palms (Arecaceae), possess characteristic silica bodies contained in well-defined cells, either in the epidermis (e.g. in grasses: Fig. 4.3B) or in vascular bundle sheath cells (e.g. in palms and orchids).

1.4 Secretory Ducts and Laticifers

In many plants, substances such as oils, resins and mucilage are secreted internally, often into specialized ducts formed either by

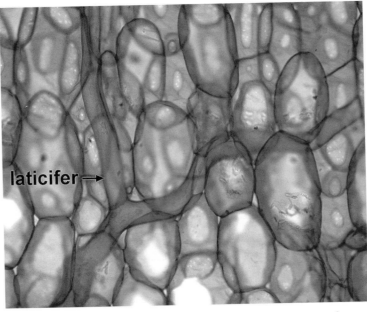

Figure 1.6 *Euphorbia eyassiana* (Euphorbiaceae), longitudinal section of stem showing branched non-articulated laticifers in parenchyma. Scale = 50 μm.

cell wall separation (schizogenous ducts) or cell wall degradation (lysigenous ducts), or a combination of the two processes (schizolysigenous ducts)[13]. Some angiosperms, especially eudicots such as *Euphorbia* and *Ficus*, produce latex from specialized cells (laticifers) that permeate their tissues (Figs 1.4, 1.6). In *Euphorbia*, the laticifers are derived from a small group of initial cells in the cotyledonary node of the embryo; these cells are coenocytes, since they undergo repeated nuclear divisions without corresponding wall formation. They grow intrusively between cells of surrounding tissues, and often branch and eventually ramify throughout the entire plant[31,71,87,90]. Coenocytic laticifers are termed non-articulated laticifers. By contrast, laticifers of a few species (e.g. *Hevea brasiliensis*, the source of commercial rubber) undergo cell-wall formation, and thus consist of linked chains of cells; these are termed articulated laticifers. Laticifers of the opium poppy

(*Papaver somniferum*) are always associated with vascular bundles[122]; the alkaloids produced in the latex of these cells are the source of narcotic analgesics such as morphine.

1.5 Transfer Cells

Transfer cells occur at the interface between tissues; they are specialized cells that facilitate transport (absorption or secretion) of soluble substances across tissue boundaries. For example, they can occur at the junction of the megagametophyte and mega-sporophyte, in companion cells in phloem tissue (especially at the node of a stem), in root nodules, in the haustoria of parasitic plants, and in the epidermis of water plants[80]. Several cells of the embryo sac and seed, including synergids, antipodals and specialized endosperm cells, have been identified as transfer cells in different species. Transfer cells are typically characterized by numerous cell-wall ingrowths protruding into their protoplasts or those of adjacent cells; these ingrowths are sometimes visible using light microscopy. Secretory cells, such as those of glandular hairs and nectaries, also frequently possess wall ingrowths. The plasma membrane of the transfer cell follows the contour of the wall ingrowths, thus increasing the surface area.

1.6 Tissues

Simple tissues, such as parenchyma, collenchyma and scleren-chyma, consist of a single cell type, though they may be interspersed with other, isolated, cell types (idioblasts). Complex tissues consist of multiple cell types, and can be divided into three main groups: dermal tissue (epidermis), ground tissue and vascular (conducting) tissue, each distributed throughout the plant body, and often continuous between the various organs. Complex tissues often include elements of several different simple tissue types; for example, secondary xylem includes not only vascular tissue, but also parenchyma and sclerenchyma.

1.6.1 Parenchyma

Parenchyma cells are typically thin-walled and often polyhedral or otherwise variously shaped, sometimes lobed. Cells with living contents that do not fit readily into other categories are often termed parenchyma cells. They are the least specialized cells of the mature plant body and often resemble enlarged meristematic cells. Parenchyma cells may occur in primary or secondary tissues. Relatively specialized types of parenchyma include certain secretory tissues and chlorenchyma, which contains chloroplasts for photosynthesis. Parenchymatous cells may be tightly packed or may be interspersed with intercellular air spaces.

Callus tissue is a cellular proliferation that is often produced at the site of a wound by divisions in parenchyma cells that have retained the ability to divide at maturity. A single isolated callus cell can be used to artificially grow a new plant using tissue culture methods.

1.6.2 Aerenchyma

Aerenchyma is a specialized parenchymatous tissue that often occurs in aquatic plants (hydrophytes). It possesses a regular, well-developed system of large intercellular air spaces (Fig. 1.7) that facilitates internal diffusion of gases. In leaves, stems and roots of some water plants (e.g. *Hydrocharis*), aerenchyma is associated with a system of transverse septa or diaphragms that provide mechanical resistance to compression. These septa are uniseriate layers of parenchyma cells that are thicker-walled than neighbouring aerenchyma cells.

1.6.3 Collenchyma

Collenchyma consists of groups of axially elongated, tightly-packed cells with unevenly thickened walls. This tissue has a strengthening function and often occurs in the angles of young stems, or in the midribs of leaves, normally in primary ground tissue. Collenchyma cells differ from fibres in that they often retain

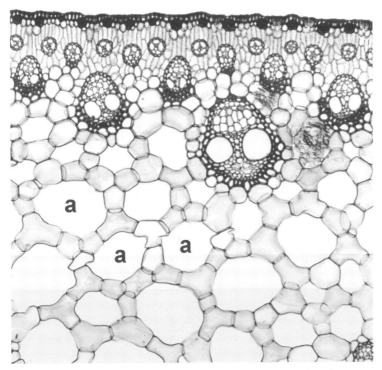

Figure 1.7 *Cyperus papyrus* (Cyperaceae), longitudinal section of leaf showing aerenchyma. a = air space. Scale = 100 µm.

their contents at maturity and do not generally have lignified walls, though they may ultimately become lignified.

1.6.4 Sclerenchyma

Sclerenchyma, also a supporting or protective tissue, consists of cells with thickened, often lignified, walls, which usually lack contents at maturity. Sclerenchyma cells occur in primary or secondary tissue, either in groups or individually as idioblasts interspersed in other tissue types. They are categorized as either fibres or sclereids, though transitional forms occur.

Fibres are long narrow cells that are elongated along the long axis of the organ concerned; they normally occur in groups. Bast fibres are extraxylary cortical fibres which can be of economic use, as in flax and hemp.

Sclereids are variously shaped and may occur throughout the plant[33]. Brachysclereids (stone cells) are isolated, approximately isodiametric cells dispersed among parenchyma cells; they develop thick secondary walls as the plant matures. Astrosclereids are highly branched cells with projections that grow intrusively into surrounding intercellular air spaces or along middle lamellae during the growth phase of the organ. Their shapes are to some extent dictated by the nature of the surrounding tissues; for example, they are often star-shaped (astrosclereids: Fig. 1.8) or bone-shaped (osteosclereids).

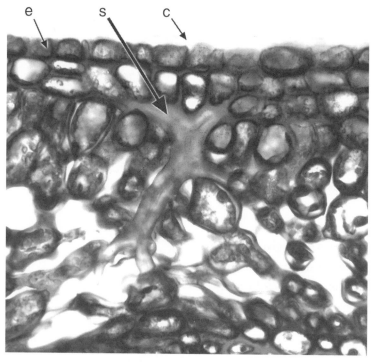

Figure 1.8 *Camellia japonica* (Theaceae), transverse section of leaf midrib showing branched sclereid(s) in ground parenchyma. c = cuticle, e = epidermis, s = sclereid. Scale = 100 μm.

1.7 Epidermis

The epidermis, the outermost (dermal) cell layer, is a complex tissue that covers the entire plant surface. The epidermis is a primary tissue derived from the outermost layer of the apical meristem. It includes many specialized cell types, such as root hairs (chapter 3.4), stomata, trichomes and secretory tissues such as nectaries, both floral and extrafloral (chapters 4.4, 5.10). The aerial plant surface is covered with a non-cellular cuticle and sometimes with epicuticular waxes (chapter 4.3.4). Undifferentiated epidermal cells are termed pavement cells. In a developing plant the protodermal cells may give rise to trichomes, stomata or pavement cells, depending on their relative position. In *Arabidopsis* most stomata develop over the junction between underlying cortical or mesophyll cells, and most root hairs develop over the junction of hypodermal cells[52].

In growing organs, anticlinal divisions (at right angles to the surface) may occur in mature epidermal cells to accommodate stem or root thickening. In older stems and roots the epidermis often splits and peels away following an increase in thickness, and is replaced by a periderm (chapter 2.9; Fig. 2.15). In some roots the epidermis is worn away by friction with soil particles, and is replaced by an exodermis, which is formed by cell-wall thickening in the outer cortical layers.

1.7.1 Stomata

Stomata are specialized pores in the epidermis through which gaseous exchange (water release and carbon dioxide uptake) takes place. They occur on most plant surfaces above ground, especially on green photosynthetic stems and leaves, but also on floral parts. Each stoma consists of two guard cells surrounding a central pore (Fig. 1.9). Cuticular ridges extend over or under the pore from the outer or inner edges of the adjacent guard cell walls. Guard cells (Fig. 4.3) are either kidney-shaped (in most plants) or dumbbell-shaped (in Poaceae and Cyperaceae). Stomata

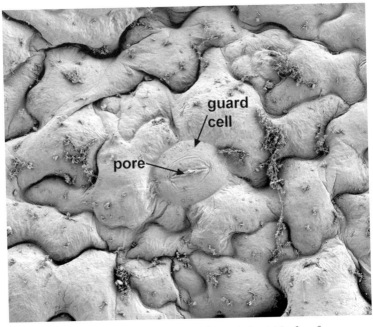

Figure 1.9 *Arabidopsis thaliana* (Brassicaceae), SEM abaxial leaf surface, showing a single stomatal pore. Scale = 10 μm.

may be sunken or raised, and are often associated with a substomatal cavity in the mesophyll, which is caused by differential expansion between the guard cell mother cell and the developing underlying mesophyll cells[42].

The epidermal cells immediately adjacent to the guard cells are termed subsidiary cells if they differ morphologically from surrounding epidermal cells. Classifications of stomatal types are based either on the arrangement of mature subsidiary cells, or on their patterns of development. Types of mature stomata include anomocytic, anisocytic, diacytic and paracytic[124]. Anomocytic stomata lack subsidiary cells entirely; anisocytic stomata possess three unequal subsidiary cells; diacytic stomata possess one or more pairs of subsidiary cells with their common walls at right angles to the guard cells; and paracytic stomata possess one or more subsidiary cells at either side of the guard cells. However, different developmental pathways may lead to similar stomatal types, so this classification could group types that are non-homologous.

Ontogenetic stomatal types include agenous, mesogenous and perigenous[85,112]. During development, a protodermal cell undergoes an unequal mitotic division to produce a larger daughter cell and a meristemoid (guard cell mother cell). In the agenous developmental type the meristemoids give rise directly to the guard cells, and there are no subsidiary cells. In the perigenous type, the meristemoid gives rise directly to the guard cells, and subsidiary cells are formed from neighbouring cells, often by oblique divisions. In the mesogenous developmental type the guard cells and subsidiary cells have a common origin; the meristemoid undergoes a further mitotic division into two cells, of which one further subdivides to form the guard cells, and the other usually forms one or more subsidiary cells. In mesoperigenous stomatal complexes the guard cells and subsidiary cells are of mixed origin. Subsidiary cells derived from the meristemoid are termed mesogene cells, whereas those derived from neighbouring cells are termed perigene cells, though in some cases mesogene cells are not distinct from surrounding epidermal cells at maturity.

1.7.2 Trichomes

Trichomes are epidermal outgrowths that occur on all parts of the plant surface (Fig. 1.10). They vary widely in both form and function, and include unicellular or multicellular, branched or unbranched forms, and also scales, glandular (secretory) hairs, hooked hairs and stinging hairs. Papillae are generally smaller than trichomes and unicellular, though the distinction is not always clear. In cases where there are several small outgrowths on each epidermal cell, these outgrowths are termed papillae, but where there is only one unicellular outgrowth per cell, the distinction is dependent on size.

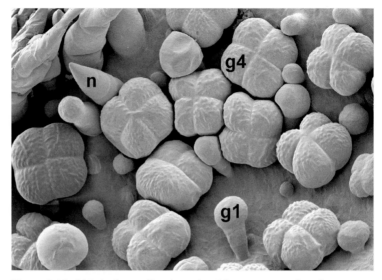

Figure 1.10 *Salvia involucrata* (Lamiaceae), trichomes on petal surface.
n = nonglandular trichome, g1 = glandular trichome with unicellular head,
g4 = glandular trichome with four-celled head. Scale = 50 μm.

Glandular trichomes usually possess a unicellular or multicellular stalk and a secretory head with one to several cells.
Secreted substances such as volatile oils collect between the
secretory cells and a raised cuticle, which later breaks to release
them. There are many different types of glandular hair, and they
secrete a variety of substances, including essential oils and salt;
some carnivorous plants' digestive juices contain proteolytic
enzymes[31]. Leaf glandular hairs of *Cannabis sativa* secrete a resinous
substance containing the mild hallucinogen tetrahydrocannabinol.
Glandular hairs of *Drosophyllum* and *Drosera* secrete both sticky
mucilage and proteolytic enzymes. The stinging hairs of *Urtica
dioica* (stinging nettle) are rigid, hollow structures that contain
a poisonous substance (Fig. 1.11). The spherical tip of the hair
is readily broken off in contact with an outside body, and the
remaining sharp point may then penetrate the skin and release

Figure 1.11 (left) Urtica dioica (Urticaceae), intact tip of stinging hair. Scale = 10 μm.

the fluid. Other examples of specialized hair types include water-absorptive leaf scales in many Bromeliaceae, and salt-secreting glands of species of Avicennia[60].

1.8 Ground Tissue

Ground tissue, sometimes termed packing tissue, forms the bulk of primary plant tissue and occupies the areas that are not taken up by vascular tissue or cavities. It has a mechanical function, and may be concerned with storage or photosynthesis. Ground tissue typically consists of parenchyma, sclerenchyma or collen-chyma, and is often interspersed with idioblasts and secretory cells or canals. Ground tissue is initially formed at the apical meristems but may be supplemented by intercalary growth, and in monocots by tissues differentiated from primary and secondary thickening meristems. In dicots the ground tissue of secondary xylem (wood), formed by the vascular cambium, consists of fibres and axial parenchyma (chapter 2.6). In older stems the central area of ground tissue (pith) often breaks down, leaving a cavity (Fig. 2.2).

1.9 Vascular Tissue

Vascular tissue consists of xylem and phloem, and may be primary or secondary in origin. Primary vascular tissue is derived from procambium, itself produced by the apical meristems, and also by the primary thickening meristem in stems of monocots (chapter 2.8). Secondary vascular tissue is derived from the vascular cambium in dicots, and from the secondary thickening meristem in a few monocots (Fig. 2.13). Both xylem and phloem are complex tissues, composed of many different cell types. Xylem is primarily concerned with water transport and phloem with food transport. Distribution of vascular tissue varies considerably between different organs and taxa.

1.9.1 Xylem

The primary function of xylem is as a water-conducting tissue. Xylem is a complex tissue composed of several cell types. The water-conducting cells are termed tracheary elements, and are typically linked to form axial chains (vessels). They have thickened lignified cell walls and lack contents at maturity. Two basic types of tracheary element can be recognized: tracheids and vessel elements; an evolutionary series from tracheids to vessel elements is widely recognized[7]. Vessel elements possess large perforations in their end walls adjoining other vessel elements, whereas tracheids lack these perforations. The perforations may have one opening (simple perforation plate) or several openings which are divided either by a series of parallel bars (scalariform perforation plate: Fig. 2.7) or by a reticulate mesh (reticulate perforation plate). In some species tracheary elements possess wall thickenings (Fig. 2.8) that are arranged either in a series of rings (annular rings), helically or in a scalariform or reticulate mesh. Annular and helical thickenings are the types most commonly found in the first-formed (protoxylem) elements. Later-formed primary tracheary elements (metaxylem) and also secondary tracheary elements typically possess bordered pits in their lateral walls.

These pits vary considerably in size, shape and arrangement; they may be oval, polygonal or elongated (scalariform pitting), organized in transverse rows (opposite pitting) or in a tightly packed arrangement (alternate pitting).

1.9.2 Phloem

Phloem has complex roles in translocation and messaging within the plant. Primary phloem is formed by the apical meristem and secondary phloem by the vascular cambium. Phloem may develop precociously in regions that require a copious supply of nutrients, such as developing sporogenous tissue.

Phloem is a complex tissue that consists of conducting cells (sieve elements) and associated specialized parenchyma cells (companion cells) (Figs. 1.12; 1.13); these two closely inter-dependent cell types are produced from a common parent cell but develop differently. Angiosperm sieve elements lack nuclei and most organelles at maturity, but retain plastids and phloem-specific

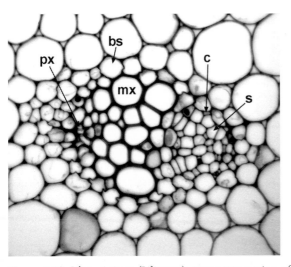

Figure 1.12 *Lilium tigrinum* (Liliaceae), transverse section of stem vascular bundle. bs = bundle sheath, c = companion cell, mx = metaxylem vessel, px = protoxylem vessel, s = sieve tube element. Scale = 100 μm.

proteins (P-proteins) which occur in several morphological forms (amorphous, filamentous, tubular and crystalline) that are often highly characteristic for particular plant families, and thus of systematic and evolutionary value[14,116]. Sieve-element plastids are classified according to their inclusions: starch (S-type plastids), protein (P-type plastids), or both. By contrast, companion cells are densely cytoplasmic, retaining nuclei and many active mitochondria.

Sieve elements are linked axially to form sieve tubes. The two basic types of sieve element, sieve cells and sieve-tube elements, are differentiated by their pore structure; most angiosperms exclusively possess sieve-tube elements. The walls of sieve elements are thin and possess characteristic regions (sieve areas) that connect adjacent sieve elements; sieve areas consist of groups

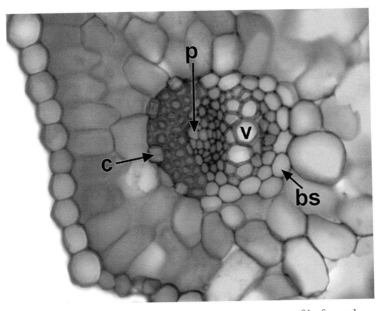

Figure 1.13 *Crocus cancellatus* (Iridaceae), transverse section of leaf vascular bundle. bs = bundle sheath, c = crystal, v = metaxylem vessel, p = phloem. Scale = 50 μm.

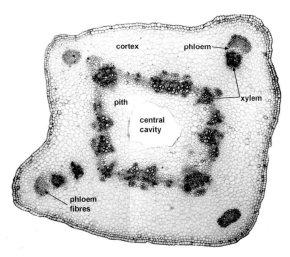

Figure 2.2 *Vicia faba* (Fabaceae), transverse section of stem. Scale = 100 μm.

fibres and sclereids, and parenchyma cells frequently become lignified as the plant ages. Ridged or angled stems often possess strengthening collenchyma at the angles, immediately within the epidermis. Many stems are photosynthetic organs with a chlorenchymatous cortex, particularly in leafless (apophyllous) plants, which normally occur in nutrient-poor habitats.

Some plant stems possess secretory cells or ducts in the ground tissue. For example, many species of *Euphorbia* possess branched networks of laticifers in the cortex (Fig. 1.6), which extend throughout the ground tissue of the stem and leaves. Plants with succulent stems, such as many Cactaceae, typically possess regions of large thin-walled cells that contain a high proportion of water. Some stems (e.g. corms of *Crocus*) are specialized as storage or perennating organs; they store food reserves in the form of starch granules, most commonly in the inner cortex. Sometimes the layer of cortical cells immediately adjacent to the vascular tissue is distinct from the rest of the cortex, and may be packed with starch granules; this is termed a starch sheath, or sometimes an endodermoid layer or endodermis, though the component cells usually lack the Casparian thickenings that are typically found in the root endodermis (chapter 3.5).

2.3 Primary Vascular System

The primary vascular system is mostly derived from the pro-cambium near the shoot apex. Primary vascular bundles possess both xylem and phloem, arranged either adjacent to each other (in collateral vascular bundles: Fig. 1.12), or with strands of phloem on both sides of the xylem (bicollateral vascular bundles), or with xylem surrounding the phloem (amphivasal vascular bundles). In woody angiosperms, internodal stem vasculature is typically arranged either in a continuous cylinder, or in a cylinder of separate or fused collateral bundles, with the phloem external to the xylem (Fig. 2.2). In some stems the bundles may be bicollateral; for example in species of *Cucurbita* internal phloem is present in addition to the external phloem. The vascular cambium, which produces secondary vascular tissue in woody species, is initially situated between the xylem and phloem within vascular bundles, but eventually extends between the vascular bundles to form a complete vascular cylinder. Some stems also possess cortical or medullary (pith) bundles, which can be associated with leaf vasculature.

In monocots, which lack a vascular cambium, the stem vascular bundles are typically scattered throughout the central ground tissue (Fig. 2.3), or sometimes arranged in two or more distinct rings. Vascular bundles may be collateral, bicollateral or amphivasal. Cortex and pith are frequently indistinct from each other, though the cortex may be defined by an endodermoid layer, or a distinct ring of vascular bundles, or in some stems, particularly inflorescence axes, by a cylinder of sclerenchyma that encloses the majority of vascular bundles. The monocot vascular system is often extremely complex[128]. Each major bundle, when traced on an upward course from any point in the stem, branches or forms

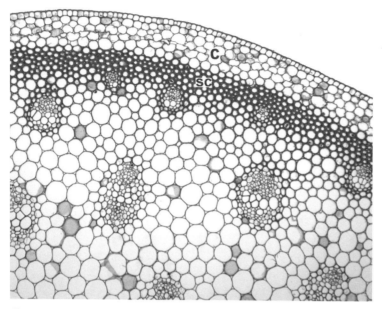

Figure 2.3 Monocot stem anatomy: *Lilium tigrinum* (Liliaceae), transverse section of inflorescence axis, showing cortex (c), surrounding central region with numerous distinct vascular bundles. sc = sclerenchymatous layer. Scale = 100 μm.

bridges with other bundles at several points before passing into a leaf. One of its major branches then continues a similar upward course towards the apex. Some palms possess literally thousands of vascular bundles in a single transverse section of the stem, though in most other monocots the number is much smaller.

2.4 Nodal Vasculature

At regions of leaf insertion on the stem (nodes), the vasculature of the leaf and stem are connected. Openings (termed lacunae or leaf gaps) occur in the stem vascular cylinder beneath their point of contact (Fig. 2.4). In eudicots and magnoliids, nodal anatomy is often characteristic of taxonomic groups, particularly the number and arrangement of leaf traces and leaf gaps.

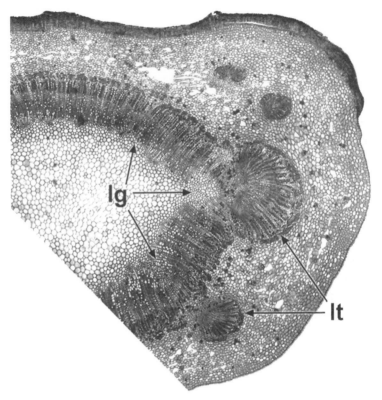

Figure 2.4 *Prunus lusitanica* (Rosaceae), transverse section of stem at node, showing connection of petiole vasculature to main vascular cylinder of stem. lg = leaf gap, lt = leaf trace. Scale = 100 μm.

Nodes may be unilacunar, trilacunar or multilacunar, depending on the number of leaf gaps in the stem vascular cylinder. This feature is most obvious in stems in which there is otherwise a continuous vascular cylinder, especially where a limited amount of secondary thickening has taken place; as a result, nodal anatomy has been studied far more extensively in woody than herbaceous plants[54]. Sometimes the number of leaf gaps per node varies within a species or individual, usually increasing with increased plant size and age[67].

The number of leaf traces departing from each gap is also generally characteristic of a species, but may vary within a plant, especially in species with unilacunar and trilacunar nodes. For example, in *Clerodendrum* two traces typically depart from a single gap, and in *Prunus* a single trace departs from each of three gaps in the central vascular cylinder (Fig. 2.4). In *Quercus* up to five traces depart through a trilacunar node. Normally, leaf trace bundles are initiated acropetally from the stem procambial system near the shoot apex, to serve developing primordia[67]. However, in some species (e.g. *Populus deltoides*) subsidiary vascular bundles are initiated at the base of each developing primordium, and grow basipetally to meet the stem procambial trace.

Nodal vasculature is further complicated by the axillary bud vascular traces, which are connected to the main stem vasculature immediately above the leaf gaps. In most species two traces diverge to supply each bud or branch.

In large woody trees, the junction of the trunk and its branches is characterized by a complex arrangement of secondary vascular tissue, which typically forms a collar around the base of the branch[99]. This branch collar is enveloped by a trunk collar, which links the vascular tissue of the trunk above and below the branch. There is no direct connection of xylem from the trunk above a branch into the branch xylem, as the tissues are oriented perpendicular to each other. If a branch dies, a protection zone forms around its base to prevent spread of infection into the trunk, and the branch is often shed.

2.5 Vascular Cambium

Increase in height, achieved by growth at the apical meristem, is inevitably followed by at least some degree of increase in stem thickness. This is achieved by different types of meristems in different species. In woody eudicots and most magnoliids (but not monocots), secondary vascular tissue (both xylem and phloem)

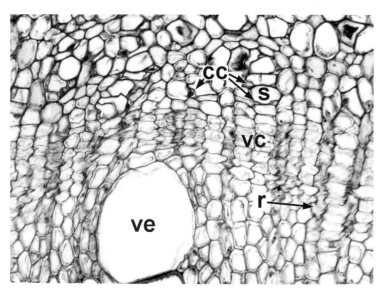

Figure 2.5 *Prunus communis* (Rosaceae). Transverse section of stem in region of vascular cambium, with secondary phloem (above) and secondary xylem (below). cc = companion cell, r = ray, s = sieve element, vc = vascular cambium, ve = vessel element.

is produced by the vascular cambium (Fig. 2.5), which usually becomes active at a short distance behind the stem apex. The vascular cambium is initiated between xylem and phloem within vascular bundles, but soon consists of an unbroken cylinder of meristematic cells. It typically generates secondary xylem (wood) at its inner edge and secondary phloem at its outer edge, though plants with anomalous secondary growth do not always follow this pattern. The amount of secondary vascular tissue produced is extremely variable, depending on the habit of the plant. Vascular cambium is absent in monocots and some herbaceous eudicots (e.g. *Ranunculus*) and magnoliids (e.g. *Saururus*).

The vascular cambium is a single cell layer (uniseriate) or several cell layers (multiseriate) if xylem and phloem mother cells are included[55]. It is a complex tissue consisting of both fusiform initials and ray initials, which form the axial and radial systems respectively. Both fusiform and ray initials are vacuolate (unlike most meristematic tissue) and plastid-rich. Fusiform initials are axially elongated cells with tapering ends. They divide periclinally to form the axial elements of secondary tissues: tracheary elements, fibres and axial parenchyma in secondary xylem, and sieve elements, companion cells and fibres in secondary phloem. Ray initials are isodiametric cells that divide periclinally to form ray parenchyma cells in both xylem and phloem. Fusiform initials sometimes give rise to new ray initials as the stem increases in circumference and new rays are formed.

2.6 Secondary Xylem

Secondary xylem (wood) varies considerably between species. The texture and density of a particular type of wood depend on the size, shape and arrangement of its constituent cells[73]. Wood is composed of a matrix of cells (Fig. 2.6), some arranged parallel to the long axis (fibres, vessels and chains of axial parenchyma cells), and others (ray parenchyma cells) forming the wood rays that extend radially from the vascular cambium towards the pith. The precise cellular arrangement in wood is often characteristic of species or genera. To observe their structure, woods are sectioned transversely (transverse section: TS) and in two longitudinal planes: along the rays (radial longitudinal section: RLS) and perpendicular to the rays (tangential longitudinal section: TLS). In some woods the vessels are solitary when viewed in transverse section (Figs 2.6, 2.10), but in other woods they are arranged in clusters or radial chains (Fig. 2.9). Axial parenchyma cells may be independent of the vessels (apotracheal) or associated with them (paratracheal), and sometimes occur in regular tangential bands. The relative abundance of axial parenchyma varies in different species, from sparse (or even completely absent) to rare cases such as *Ochroma pyramidale* (balsa), in which

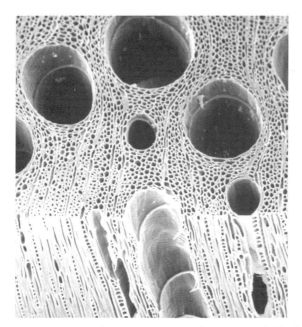

Figure 2.6 Secondary xylem: *Quercus robur* (Fagaceae), block of wood at edge of transverse and tangential longitudinal surfaces, showing large early (spring) wood vessels.

axial parenchyma cells are often more abundant than fibres, making this type of wood soft and easy to carve.

Rays are termed uniseriate if they are one cell wide tangentially, and multiseriate if they are more than one cell wide, viewed in TS and TLS. Sometimes both uniseriate and multiseriate rays occur in the same wood, as in *Quercus*. Ray cells vary in shape (best viewed in RLS); homocellular rays are composed of cells of similar shapes, whereas in heterocellular rays the cells are of different shapes. Other aspects of variation in the structure of hardwoods include the presence of either axial or radial secretory canals in some woods (Fig. 2.10), the storied (stratified) appearance of various elements, particularly rays, or the fine structure of the vessel walls (intervascular pitting, perforation plates and wall thickenings: chapter 1.7.1). For example, in *Tilia cordata* (Fig. 2.8), the vessel

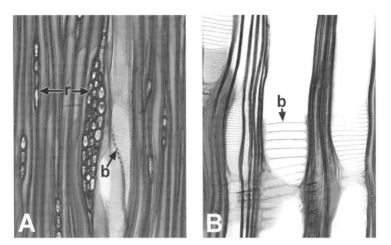

Figure 2.7 *Betula utilis* (Betulaceae). Wood in (A) tangential longitudinal section (TLS) and (B) radial longitudinal section (RLS). b = bar of scalariform perforation plate, r = ray. Scale = 100 μm.

element walls are helically thickened, and in many Fabaceae the pit apertures are surrounded by numerous warty protuberances, termed vesturing[19]. Perforated ray cells, an unusual feature of some woods, are ray cells that link two vessel elements and themselves resemble and function as vessel elements, with perforation plates corresponding to those of the adjacent vessel elements. However, like other ray cells, perforated ray cells are formed from ray initials rather than from fusiform initials, like vessel elements.

In many woody temperate plants cambial activity is seasonal (usually annual), which results in the formation of growth rings. The secondary xylem formed in the early part of the season (early wood or spring wood) is generally less dense and consists of thinner-walled cells than the xylem formed later in the growing season (late wood or summer wood). In ring-porous woods the vessels are considerably larger in early wood than in late wood (Fig. 2.11). In diffuse porous woods the main distinction between early and late wood is in size and wall thickness of the fibres (Fig. 2.9). As woody plants age and their trunks increase in

Figure 2.8 *Tilia olivieri* (Tiliaceae), SEM inside surface of vessel element showing wall thickenings and intervascular pitting.

girth, the central area becomes non-functional with respect to water transport or food storage, and the vessels frequently become blocked by tyloses. Tyloses are formed when adjacent parenchyma cells grow into the vessels through common pit fields. The central non-functional area of the trunk, the heartwood, is generally darker than the outer living sapwood.

In some woody angiosperms, particularly climbing plants (lianas) such as many Bignoniaceae (Fig. 2.12), secondary growth does not fit the "normal" pattern of xylem and phloem production, and is termed anomalous secondary growth. For example, some plants develop regions of phloem (included or interxylary phloem) embedded in the xylem, either in islands (e.g. in *Avicennia*) or in alternating concentric bands. Other examples have irregularly divided or deeply fissured areas of xylem and phloem, or stems that are flattened or otherwise

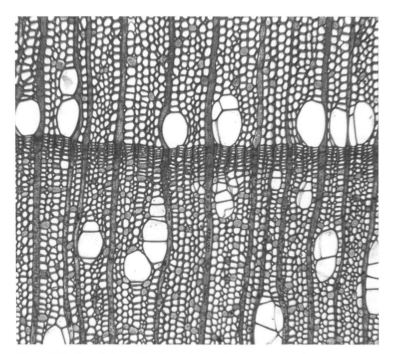

Figure 2.9 *Alnus glutinosa* (Betulaceae), wood, transverse section. Scale = 100 μm.

irregularly shaped[73]. Such anomalous forms are achieved either by the formation of new vascular cambia in unusual positions or by the unusual behaviour of the existing cambium in producing phloem instead of xylem at certain points.

2.7 Secondary Phloem

Secondary phloem is also a product of the vascular cambium in woody species. As in secondary xylem, secondary phloem consists of both axial and radial systems, formed from the fusiform and ray initials respectively. Phloem rays are radially continuous with xylem rays, and may be similarly uniseriate or multiseriate, though in transverse section they often appear dilated towards the cortex as a result of cell divisions to accommodate increase in stem thickness (Fig. 2.11). At their outer periphery, the

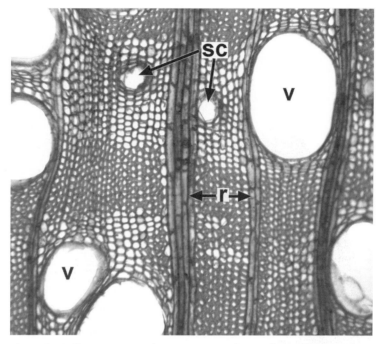

Figure 2.10 *Shorea resina-nigra* (Dipterocarpaceae), wood, transverse section showing vessels (v) and axial secretory canals (sc), r = ray. Scale = 100 μm.

parenchymatous ray cells are often difficult to distinguish from cortical cells. Older ray cells sometimes become lignified to form sclereids. The axial system of the phloem consists of sieve elements and companion cells, as in primary phloem (chapter 1.9.2). It also typically includes fibres, sclereids and axial parenchyma cells. In some species fibres are formed in groups at regular intervals, resulting in characteristic tangential bands of fibres alternating with groups of sieve elements and parenchyma cells.

2.8 Primary and Secondary Thickening Meristems

In monocots, which lack a vascular cambium, increase in stem diameter is typically relatively limited. However, most monocots

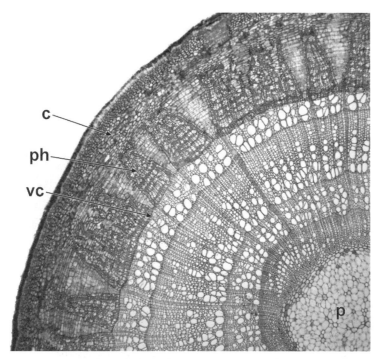

Figure 2.11 *Tilia olivieri* (Tiliaceae), transverse section of twig with slightly more than three years growth, c = cortex, p = pith, ph = phloem, vc = vascular cambium. Scale = 100 μm.

and a few other thick-stemmed angiosperms, especially species with short internodes and crowded leaves, possess a primary thickening meristem (PTM) near the vegetative shoot apex[88]. The PTM (Fig. 2.13) is situated in the pericyclic region. It consists of a narrow multiseriate zone of meristematic cells that produces radial derivatives, usually a limited amount of parenchyma towards the outside (centrifugally), and both parenchyma and discrete vascular bundles towards the inside (centripetally). In addition to primary stem thickening, the PTM is responsible for formation of linkages between root, stem and leaf vasculature. Also, it frequently retains meristematic potential further down

Figure 2.12 *Tynanthus elegans* (Bignoniaceae). Transverse section of woody stem showing anomalous secondary growth: xylem region with four deep fissures of phloem. Scale = 1 mm.

the stem and is the site of adventitious root production in some species.

The PTM normally ceases activity at a short distance behind the apex, and subsequent stem thickening is limited. Tree-forming palms possess an extensive PTM that forms a large sunken apex; considerable further stem thickening occurs by subsequent division and enlargement of ground parenchyma cells. This is termed diffuse secondary growth.

In some woody monocots in the order Asparagales (e.g. *Agave*, *Aloe*, *Cordyline*, *Yucca*) further increase in stem thickness is achieved by means of a secondary thickening meristem (STM) (Fig. 2.14). The STM is essentially similar to the PTM in that it is located in the pericyclic region and produces radial derivatives. However, it is active further from the primary apex and produces second-ary vascular bundles that are often amphivasal and radially elongated. In some species (e.g. *Nolina recurvata*, *Cordyline terminalis*)

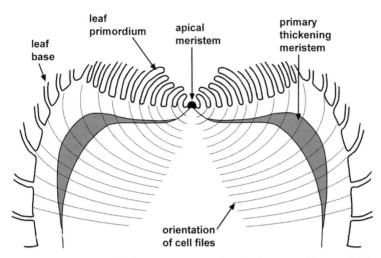

Figure 2.13 Primary thickening meristem (PTM): diagram of longitudinal section of the crown of a typical thick-stemmed monocot, showing orientation and extent of radial PTM derivatives. Vascular strands not shown. (Adapted from DeMason 1983).

the PTM and STM are axially discontinuous[104,105], whereas in others (e.g. *Yucca whipplei*) they are axially continuous[26]. Apart from the distance from the apex, there are no precise criteria for distinguishing between derivatives of the two meristems, and transitional forms exist. Thus, they are perhaps best regarded as developmental phases of the same meristem.

The PTM and STM are not homologous with the vascular cambium, because the vascular derivatives are arranged in different ways. The vascular cambium produces phloem centrifugally and xylem centripetally, whereas most derivatives of the PTM and STM are centripetal, and consist of a parenchymatous ground tissue and discrete vascular bundles containing both xylem and phloem. Furthermore, the PTM originates in ground tissue, is a tiered meristem, and is often fairly diffuse, especially near the shoot apex (Fig. 2.13). By contrast the vascular cambium is typically uniseriate and initially originates within vascular tissue, though it later extends between bundles.

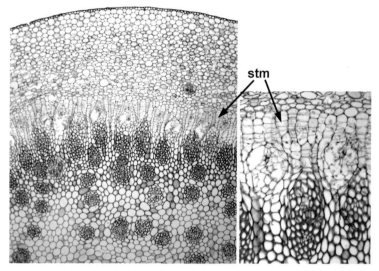

Figure 2.14 Secondary thickening in monocots: *Dracaena indivisa* (Ruscaceae), transverse section of stem showing secondary thickening meristem (STM) and radial internal vascular derivatives. Scale = 100 μm (left hand image).

2.9 Periderm

Periderm is a protective tissue of corky (suberinized) cells that is produced either as a response to wounding or in the outer layers of the cortex of a stem or root that has increased in thickness. The periderm consists of up to three layers: phellogen, phellem and phelloderm. The phellogen is a uniseriate meristematic layer of thin-walled cells that produces phellem to the outside, and (in some cases) phelloderm to the inside. The phellem cells constitute the corky tissue. They are tightly-packed cells that lack contents at maturity. They possess deposits of suberin and sometimes lignin in their walls, and form an impervious layer to prevent water loss and protect against injury. Phelloderm cells are non-suberinized and parenchymatous, and contribute to the secondary cortex.

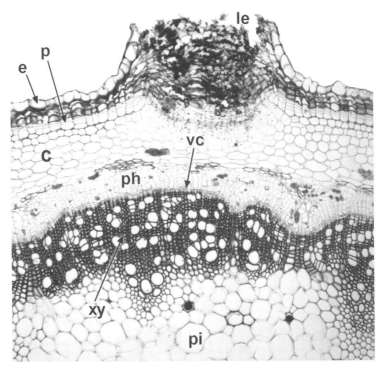

Figure 2.15 *Sambucus nigra* (Caprifoliaceae). Transverse section of stem surface, showing periderm forming in outer cortical layers. c = cortex, e = epidermis, le = lenticel, p = periderm, ph = secondary phloem, pi = pith, vc = vascular cambium, xy = secondary xylem. Scale = 100 μm.

A periderm commonly occurs in the cortex of secondarily thickened stems, to replace the epidermis, which splits and peels away (Fig. 2.15). The phellogen may originate either adjacent to the epidermis (or even within the epidermis) or deeper in the cortex. Sometimes several phellogens form almost simultaneously. The pattern of periderm formation largely dictates the appearance of the bark of a woody plant. For example, the smooth papery bark of a young silver birch tree (*Betula pendula*) is formed because the periderm initially expands tangentially with the increase in stem diameter, but later flakes off in thin papery sheets as abscission bands of thin-walled cells are formed. In the trunk of cork oak (*Quercus suber*), the initial phellogen may continue activity indefinitely, and produces seasonal growth rings. In the commercial process it is removed after about 20 years to make way for a second, more vigorous phellogen, which produces the commercial cork.

Many species possess lenticels in the bark (Fig. 2.15); these are areas of loose cells in the periderm, which are often initially formed beneath stomata in the epidermis, and are thought to be similarly concerned with gaseous exchange.

3

Root

3.1 Primary Root Structure

The seedling radicle ultimately becomes the primary root (tap root), which frequently develops side branches (lateral roots). In monocots the seedling radicle commonly dies at an early stage; the stem-borne (adventitious) roots of the mature plant originate from differentiated cells (Fig. 3.4). Adventitious roots can be branched or unbranched. Although roots can originate from various organs, their basic primary structure retains a characteristic root groundplan that is different from that of the stem. Each root possesses clearly-defined concentric tissue regions: dermal tissue (epidermis), ground tissue (cortex, including the endodermis) and central vascular tissue surrounded by a pericycle (Fig. 3.3).

3.2 Root Apex

Root apices possess a terminal protective root cap and a proximal root apical meristem[8,32] (Fig. 3.1). The quiescent centre is a group of relatively inactive cells at the very centre and tip of the root apical meristem. The cells of the quiescent centre divide infrequently; their role is obscure, but they maintain initial cells in an undifferentiated state. These cells, together with the root cap initials, are derived from the uppermost cell of the suspensor (hypophysis) in the embryo[18] (Fig. 6.7). Cell division activity occurs in the cells surrounding the quiescent centre.

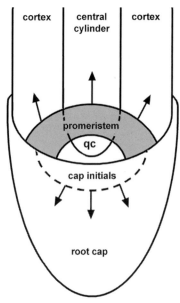

Figure 3.1 Diagram of root apical organization in *Zea mays* (Poaceae), a species with closed structure. Arrows indicate direction of displacement of cell derivatives. (Adapted from Feldman 1984).

In *Arabidopsis thaliana* the initial cells lie in clearly defined regions relative to the quiescent centre, the pericycle and vascular initials proximal to it (on the shoot side), the root cap and epidermis initials distal to it (on the root cap side) and the cortical and endodermal initials radial to it. However, in other species (e.g. *Vicia faba*) there is an undifferentiated initiating region common to all root tissues[102]. The active region is termed the promeristem.

The junction between the root cap and the root apical meristem is either clearly defined by a distinct cell boundary (termed closed organization, as in *Zea mays* and *Arabidopsis thaliana*), or ill-defined (termed open structure, as in *Vicia faba*: Fig. 3.2), though intermediates exist (e.g. in *Daucus carota*)[11,21]. In open meristems the boundary between the cap and the rest of the root is unstable.

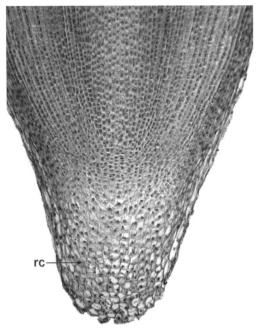

Figure 3.2 *Vicia faba* (Fabaceae), longitudinal section of root apex, showing open apical structure. rc = root cap. Scale = 100 μm.

3.3 Root Cap

The root cap is composed of several layers of parenchymatous cells. The cells of the root cap are initially derived from the apical meristem. However, ontogenetic studies in maize (*Zea mays*), a species with "closed" root apical structure (Fig. 3.1), have shown that the cap initials become established and independent from the apical meristem at an early stage in seedling develop-ment[8]. The cap meristematic cells, located adjacent (distal) to the quiescent centre, produce derivatives that are eventually displaced towards the outside of the root cap, and subsequently sloughed off, contributing to the external slime that allows the root to push through the soil. Cells are generated and lost in the root cap at approximately the same rate.

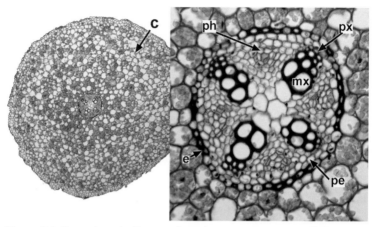

Figure 3.3 *Ranunculus acris* (Ranunculaceae), transverse section of root, with detail of central vascular region (inset). c = cortex, e = endodermis, mx = metaxylem, pe = pericycle, ph = phloem, px = protoxylem. Scale = 100 μm.

3.4 Root Epidermis and Hypodermis

In root apices with closed organization, the root epidermis is associated either with cortical cells (in most monocots) or with cap initials (in most other angiosperms); in root apices with open organization the precise origin of the root epidermis is relatively difficult to determine[21].

In eudicots the root epidermis (rhizodermis) is typically uniseriate, as in other parts of the plant. In monocots the root epidermis is normally persistent and remains as the outermost layer of the root. A velamen is particularly characteristic of aerial roots of Orchidaceae and Araceae. Velamen cells of a mature root are dead, and often become saturated with water for storage purposes, whereas a persistent rhizodermis consists of living cells. A velamen is usually multilayered but can also be one-layered; the cell walls are often partly thickened and sometimes lignified. In Orchidaceae, velamen cells frequently possess wall striations.

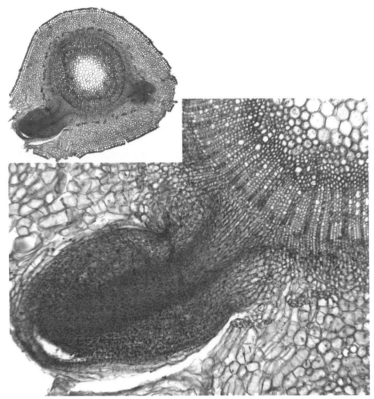

Figure 3.4 *Ligustrum vulgare* (Oleaceae), transverse section of stem with adventitious roots. Scale = 100 μm.

Most angiosperms possess absorptive root hairs in underground roots, usually confined to a region about a centimetre from the root apex, beyond the meristematic region, but in an area where cells are still enlarging. In general, root hairs persist for only a limited period before withering. This region of the root is the most active in absorption of water, and the root hairs serve to present a greater surface area for this purpose. Root hairs are formed from epidermal cells by apical intrusive growth. In some plants only specialized root epidermal cells (trichoblasts) are capable of root hair production. Trichoblasts are formed in meristematic epidermal cells that overlie the junction between two cortical cells[18]. Thus, in many species the root epidermis is dimorphic and clearly differentiated into short cells (trichoblasts) and long cells (sometimes termed atrichoblasts), as in *Arabidopsis thaliana*. Some other species (including many monocots such as species of Asparagales and Araceae) instead possess a dimorphic hypodermal layer immediately below the root epidermis[63]; this is normally interpreted as the outermost cortical (exodermal) layer but may represent the innermost layer of a multilayered persistent rhizodermis. The hypodermal short cells resemble trichoblasts, and are probably transfusion cells.

3.5 Root Cortex and Endodermis

The cortex is the region between the pericycle and the epidermis, including the innermost layer, the endodermis. In underground roots the rhizodermis becomes worn away, and is replaced as an outer layer either by a periderm that forms in the cortex (in most woody eudicots and magnoliids) or by a suberinized or lignified exodermis (in some monocots), which is sometimes multilayered.

Apart from these specialized layers, most cortical cells are parenchymatous and often perform an important storage function. In some plants, such as *Daucus carota* (carrot), the tap root is a modified swollen storage organ with a wide cortex. In most roots the bulk of the cortical cells are formed by sequential periclinal divisions, the innermost cells (later the endodermis) being the last formed.

Many plants with underground stems (corms, bulbs or rhizomes), particularly bulbous or cormous monocots such as *Crocus*, *Freesia* and *Hyacinthus*, periodically produce contractile roots which draw the stem deeper into the soil[57]. These roots grow downwards, and then shorten vertically and expand radially. They are recognizable by their wrinkled surface, and characteristically possess two or three clearly distinct concentric regions of cortical parenchyma, distinguishable by cell size, including a region of collapsed outer cortical cells interspersed with occasional thicker-walled cells. In some species the process of root contraction is initiated by active cell enlargement in the inner cortex, followed by collapse of outer cortical cells and subsequent surface folding. In other species the collapse of outer cortical cells results from the difference between atmospheric pressure and relatively low xylem pressure (due to transpiration), causing centripetal loss of turgidity.

The endodermis is a uniseriate cylinder of cortical cells surrounding the central vascular region, adjacent to the pericycle. Endodermal cells are typically characterized by deposition of a band of suberin or lignin in their primary walls, termed a Casparian strip, which forms a barrier against non-selective passage of water through the endodermis. Older endodermal cells often possess thick lamellated secondary walls, in most cases on the inner periclinal wall, so that the Casparian strip is not apparent. The secondary wall is often lignified, and therefore serves as a second effective barrier to water loss. Occasional endodermal cells (passage cells) can remain thin-walled, probably for selective passage of water between the cortex and vascular region.

3.6 Pericycle and Vascular Cylinder

The vascular tissue in the centre of the root is surrounded by a single layer (or rarely, more layers) of thin-walled cells, termed the pericycle (Fig. 3.3). Both the pericycle and vascular tissue are derived from cells on the proximal (shoot) side of the quiescent centre. The pericycle is potentially meristematic in younger roots, as it is the site of lateral root initiation, but in older roots it can become lignified.

The primary vascular tissue consists of several strands of phloem alternating with the rays of a central area of xylem that appears star-shaped in transverse section. In a mature root, the protoxylem elements, which were the first-formed and are the narrowest in diameter, are located at the tips of the rays, nearest to the pericycle. The metaxylem elements are larger and located closer to the centre of the root. Both xylem and phloem are exarch in the root (i.e. they mature centripetally). Similarly, the protophloem is located close to the pericycle, in contrast with the metaphloem, which is situated closer to the centre of the root.

Roots possess two, three, four or more protoxylem poles (rays), in which case they are said to be diarch, triarch, tetrarch or polyarch respectively. There is often variation in the number of xylem poles, sometimes even within the same plant, depending on the diameter of the root. Most commonly, roots possess relatively few xylem poles (usually two, three or four) and the central region is occupied by a group or ring of xylem vessels. However, some monocots (e.g. Iris) possess polyarch roots, and the centre of the root is parenchymatous, sometimes becoming lignified in older roots.

3.7 Initiation of Lateral and Adventitious Roots

Lateral roots are branches of the tap root. They are initiated in relatively mature tissues some distance from the apex, often in acropetal sequence; the most recently-formed lateral roots are usually those nearest to the root apical meristem. In angiosperms, lateral roots have a deep-seated (endogenous) origin. Root formation is usually initiated in groups of "founder cells" in the pericycle, often adjacent to the xylem poles. The position of lateral root initiation in the pericycle is usually at a point adjacent to a protoxylem pole, unless the root is diarch, in which case initiation is sometimes opposite a phloem pole. However, in monocots lateral root initiation can be opposite either protoxylem or phloem poles, though in roots with a large number of vascular poles it is often difficult to determine the precise site of initiation[72]. The founder cells undergo a series of periclinal and anticlinal divisions to form a lateral root primordium. In many species some subsequent cell divisions occur in the endodermis, so that ultimately both the pericycle and the endodermis contribute to the tissues of the lateral root. The growing lateral root pushes its way through the cortex and epidermis of the parent root, either by mechanical or enzymatic action.

Adventitious roots are formed in other parts of the plant, primarily stem tissue. They have various sites of origin, from deep-seated (endogenous) (Fig. 3.4), to (more rarely) exogenous, arising from superficial tissues such as the epidermis (e.g. in surface-rooting *Begonia* leaves). In most monocots adventitious roots arise from cell divisions in the pericyclic region of the stem; the primary thickening meristem contributes to adventitious root formation (chapter 2.8). Adventitious roots are often formed at nodes on the stem, which is why in horticulture cuttings are most commonly taken from just below a node. Adventitious roots may also form from callus tissue at the site of a wound.

3.8 Secondary Growth in Roots

In some woody eudicots the thickening and strengthening of the root system is important in supporting the trunk. Most dicot roots possess at least a small amount of secondary thickening (Fig. 3.5), with the exception of a few herbaceous species such as *Ranunculus* (Fig. 3.3). In contrast, secondary growth in roots is extremely rare in monocots, even among arborescent or woody species that possess a secondary thickening meristem (chapter 2.8). A notable exception is *Dracaena*, in which a limited region of secondary tissue is formed[111].

As in the stem, secondary vascular tissues of the root are produced by a vascular cambium. This initially develops in the

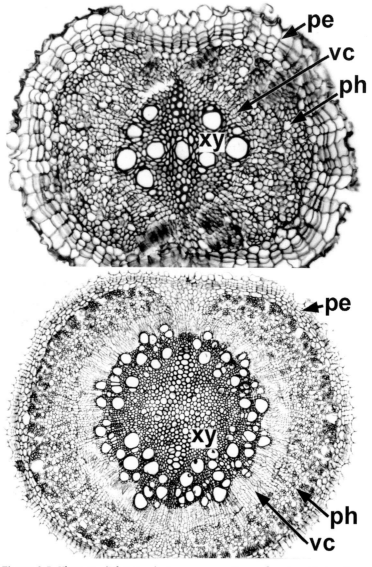

Figure 3.5 *Ulmus* sp. (Ulmaceae), transverse sections of roots with (top) recent secondary growth and (bottom) more extensive secondary thickening. pe = periderm, ph = phloem, vc = vascular cambium, xy = xylem.

regions between the primary xylem and phloem, then in derivatives of cell divisions in the pericycle next to the xylem poles. Since cambial activity proceeds in this sequence, the xylem cylinder soon appears circular in transverse section (Fig. 3.5). Further pericyclic cell divisions result in a secondary cortex, and in many cases a periderm forms, particularly where secondary growth is extensive. The epidermis splits and is sloughed off together with the primary cortex and endodermis. Root secondary xylem usually resembles that of the stem in the same plant, but may differ in several respects. For example, in *Quercus robur* stem wood is ring porous, with earlywood vessels markedly larger than latewood vessels, but root wood is diffuse porous, with vessels of relatively consistent sizes across each growth ring. As with trunk wood, root wood of individual taxa often exhibits identifiable characteristics[24].

3.9 Roots Associated with Micro-Organisms

Many vascular plants form symbiotic relationships with soil microorganisms. In legumes, nitrogen-fixing bacteria invade the root cortex through root hairs, and stimulate meristematic activity in the cortex (and sometimes also in the pericycle) to form a root nodule, which often becomes elongated to resemble a short lateral root. Other soil micro-organisms may induce the formation of modified lateral roots. For example, in many woody angiosperms, invading filamentous bacteria promote the development of short, swollen lateral roots, and in some temperate woody forest species, especially in the families Fagaceae and Betulaceae, ectomycorrhizal fungi form a mantle over stunted lateral roots. By contrast, the more common endomycorrhizal fungi, which invade the cells of the host root, often have little influence on root morphology (Fig. 3.6).

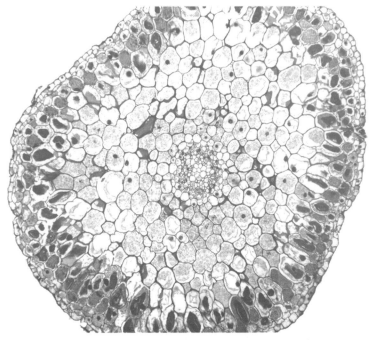

Figure 3.6 *Neottia nidus-avis* (Orchidaceae), transverse section of root with fungal hyphae in outer cortex. Scale = 100 μm.

3.10 Haustoria of Parasitic Angiosperms

Some angiosperms are parasitic on the roots, stems and leaves of other angiosperms. These include mistletoe (*Viscum album*), dodder (*Cuscuta* spp.), sandalwood (species of Santalaceae), broomrape and figwort (species of Orobanchaceae). Parasitic plants produce highly modified structures, termed haustoria, that penetrate the host tissue to transfer nutrients from the host to the parasite[66]. A primary haustorium is a direct outgrowth of the apex of the radicle of the parasite. A secondary haustorium is a lateral organ that develops from a modified adventitious root or from outgrowths of roots or stems.

In some parasitic species the haustorium penetrates the host tissue to the xylem and forms a continuous xylem

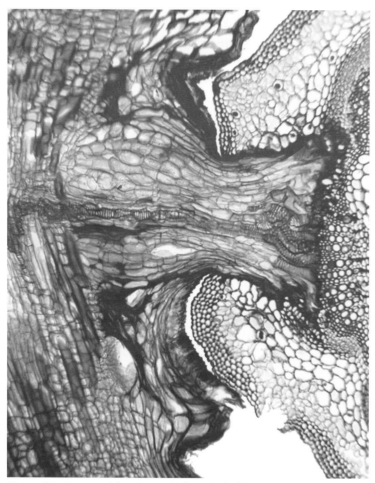

Figure 3.7 *Cuscuta cordofana* (Convolvulaceae) (left) parasitic on stem of
Trifolium (right); parasite haustorium extending into host and forming xylem
bridge. Scale = 100 μm.

bridge (Fig. 3.7). The epidermal cells in contact with the host
become elongated and secretory, and the centre of the haustorium
develops an intrusive process that grows into the host by
enzymatic and mechanical action. In many parasites (e.g. sandal-
woods), the developing haustorium forms a mantle of parenchymatous tissue around the host organ. In some Loranthaceae the

haustorium does not invade the host tissues after forming a mantle, but instead induces the host tissue to form a placenta-like outgrowth of vascular tissue to supply nutrients to the parasite. When the parasite dies, a woody outgrowth of convoluted host tissue remains as a vestige known as a "woodrose".

Leaf

4.1 Leaf Morphology and Anatomy

Angiosperm leaves display much morphological and anatomical diversity. Mature leaves of monocots are typically narrow and consist of a linear lamina with parallel venation and a leaf base that ensheathes the stem. This contrasts with the typical leaf of eudicots and magnoliids, which has a well-defined petiole and elliptical blade (lamina) with reticulate venation. However, exceptions and transitional forms are common; for example, leaves of some monocots (e.g. *Dioscorea* and *Smilax*) are petiolate and net-veined, and leaves of some eudicots (e.g. some Apiaceae) are linear. Some species possess compound leaves in which individual leaflets are borne either on a central stem-like axis (pinnate leaves; e.g. tomato, *Solanum lycopersicum*) or radiate from a single point at the distal end of the petiole (palmate leaves; e.g. *Arisaema*).

Some species that grow in dry (xeric) or seasonally dry habitats, or otherwise nutrient-deficient habitats, possess specialized xeromorphic features, including sunken stomata to minimize water loss and well-developed sclerenchyma to provide mechanical support and minimize tissue collapse. Other xeromorphic features include the presence of a hypodermis or thick epidermis and thick cuticle which diminish the intensity of light that reaches photosynthetic tissue. Well-developed palisade tissue is also sometimes correlated with high light intensity. Some xeromorphic species possess thick, sometimes even succulent, leaves; others have terete (centric or cylindrical) leaves, or hairy leaves, or even folded (plicate) or rolled leaves (Fig. 4.1). Leaf rolling or folding allows minimal water loss in unexpanded developing leaves that are still rolled or folded within the bud or superadjacent leaf, but still achieves a large surface area at maturity. Succulent plants possess large thin-walled cells for water storage. Thick and terete leaves possess a reduced surface/volume ratio that helps to reduce water loss. Other features, such as poorly-developed sclerenchyma and large air spaces in the ground tissue (aerenchyma) are often associated with water plants (hydrophytes).

The mature lamina consists of an adaxial and abaxial epidermis enclosing several layers of mesophyll cells that are interspersed

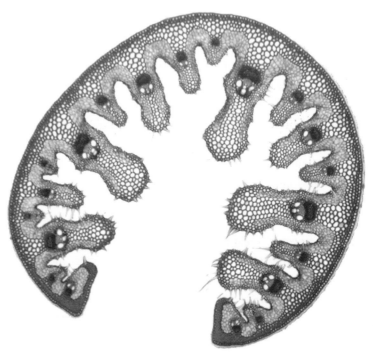

Figure 4.1 *Ammophila arenaria* (Poaceae), transverse section of rolled leaf. Scale = 500 μm.

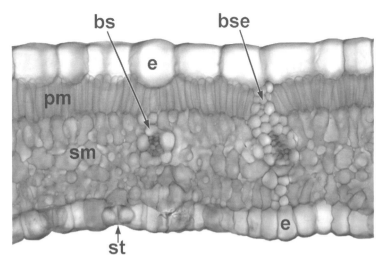

Figure 4.2 Ficus pretoriae (Moraceae), transverse section of leaf. bs = bundle sheath, bse = bundle sheath extension, e = epidermis, pm = palisade mesophyll, sm = spongy mesophyll, st = stomatal pore. Scale = 100 μm.

with a network of vascular bundles (Fig. 4.2). Each tissue may be variously differentiated into specialized cell types, though the degree of differentiation varies considerably among taxa. Leaves are most commonly bifacial (dorsiventral), in which case the upper and lower surfaces differ morphologically; for example, in relative numbers of stomata and trichomes. However, some species possess isobilateral or unifacial leaves, in which the epidermis and mesophyll are similar on both surfaces. In unifacial-leafed monocots the leaf base is bifacial and the lamina is unifacial and either flattened (e.g. in *Acorus* and most Iridaceae) or rounded (terete) (e.g. in some *Allium* species).

Stipules are (often leaf-like) foliar appendages that occur at leaf base in some angiosperms, especially eudicots. Ligules are outgrowths of the abaxial epidermis that occur in the region between the leaf sheath and lamina in grasses and several other monocot groups[94]; they are derived from a cross zone in the leaf primordium. Squamules are small hair-like structures associated with the axils of foliage leaves and sometimes floral pedicels, especially in Alismatales (e.g. *Potamogeton*) and Brassicaceae (e.g. *Arabis*); in some species they secrete mucilage.

4.2 Leaf Development

Leaves are initiated from groups of founder cells close to the stem apex. These undergo periclinal divisions either in the outermost cell layers or in the layers immediately below them, to form small conical projections (leaf primordia)[46,107]. In monocots the leaf primordium rapidly develops into a bifacial hood-like structure, and the base of the primordium partially or wholly encircles the stem, forming a leaf sheath[94]. Within a simple leaf, most subsequent meristematic activity occurs in a highly plastic transition zone between the precursor tip and sheath[46]. One or more adaxial cross (transverse) meristems in the transition zone give rise to many other structures such as ligules and stipules.

The adaxial marginal cells divide rapidly to form a flattened leaf blade. This marginal growth is suppressed in the region that later becomes the petiole, and in many monocots it often occurs at the same time as apical growth. Marginal growth is subsequently replaced by cell divisions across the whole leaf blade; by this stage the approximate number of cell layers has been established, and the whole lamina functions as a plate meristem. Cell divisions are mainly anticlinal, resulting in regular layers of cells that are disrupted only by the differentiation and maturation of the vascular bundles.

Rates of growth and cell division sometimes vary in different parts of the leaf. Individual leaflets of a compound leaf may be produced either acropetally or basipetally. For example, in the tomato (*Solanum lycopersicum*), the uppermost (distal) leaflets are initiated first, followed by middle and lower leaflets (i.e. a basipetal sequence)[56]. Smaller intermediate leaflets are formed later, in a more chaotic sequence. In some monocots, meristematic activity governing leaf elongation is restricted to a region at the base of the leaf, the basal rib meristem; this results in axial files of cells that increase in maturity towards the distal end of the leaf. The unifacial leaves of some monocots (e.g. *Acorus*), which possess a bifacial sheathing leaf base and a unifacial upper blade, result from suppressed marginal growth and the presence of an adaxial (ventral) meristem in the transition zone[61,62].

4.3 Leaf Epidermis

The leaf epidermis is a complex tissue that usually consists of a single layer of cells, though in a few species (e.g. *Ficus* and *Peperomia*) the epidermis proliferates to form several cell layers (a multiple epidermis). At maturity it is difficult to distinguish a multiple epidermis from a hypodermis. The specialized elements of the leaf epidermis are essentially the same as those of the stem: stomata, trichomes, papillae, surface sculpturing, epicuticular wax and variously differentiated pavement epidermal cells.

4.3.1 Pavement Epidermal Cells

In surface view, pavement epidermal cells can be elongated or more or less isodiametric, and their anticlinal walls can be straight or undulating (Fig. 4.3A). Anticlinal cell walls are often more sinuous on the abaxial than the adaxial surface of the same leaf. Cells that lie over veins are often elongated in the direction of the veins. In linear leaves of the type found in many monocots, epidermal cells are typically elongated parallel to the long axis of the leaf (Fig. 4.3B).

Epidermal cells frequently vary in size and wall thickness in different parts of the same leaf. In some Poaceae (e.g. *Zea mays*), enlarged cells (termed bulliform cells) occur in restricted regions

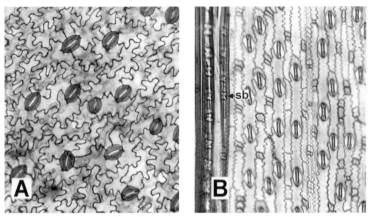

Figure 4.3 Abaxial leaf surfaces with stomata: (A) *Paeonia officinalis* (Paeoniaceae); (B) *Arundo donax* (Poaceae). sb = silica body.

of the leaf epidermis (Fig. 4.9). In some species these cells play a role in the unrolling of the leaf in response to turgor pressure and water availability. Occasional epidermal cells may contain crystals or silica bodies. Cystoliths can be restricted to individual epidermal cells (e.g. in the family Opiliaceae), or they can span both the epidermis and underlying mesophyll (e.g. in *Ficus*: Fig. 4.4). They contain a body encrusted with calcium carbonate that is attached to the cell wall by a silicified stalk[36]. Silica bodies occur in the leaf epidermis of the monocot families Cyperaceae, Poaceae (Fig. 4.3B) and Arecaceae. The grass epidermis typically consists of both long and short cells, the short cells sometimes forming the bases of trichomes.

4.3.2 Stomata

Since they control gaseous exchange (water release and carbon dioxide uptake) (chapter 1.5), stomatal pores occur in almost all angiosperm leaves, though their distribution varies in different species. In many species, stomata are present on both leaf surfaces (i.e. in amphistomatic leaves), whereas in others they

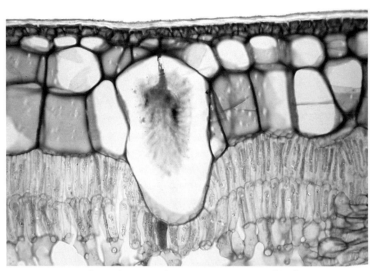

Figure 4.4 *Ficus elastica* (Moraceae), transverse section of leaf showing cystolith. Scale = 100 μm.

are restricted to the abaxial surface (hypostomatic leaves). Rarely, stomata are restricted to the adaxial surface (in epistomatic leaves); this condition occurs mainly on the floating leaves of aquatic plants. Within each leaf, stomata most frequently occur on regions overlying the chlorenchymatous mesophyll rather than the veins. In xeromorphic species stomata are often protected to restrict water loss; for example, individual stomata may be sunken, or groups of stomata may be restricted to hair-lined grooves or depressions on the abaxial leaf surface (Fig. 4.5). This creates a pocket of water vapour, and thus reduces water loss by transpiration.

4.3.3 Trichomes and Papillae
Similarly, the distribution of leaf-borne trichomes (chapter 1.7.2) varies among species. Trichomes can occur on the entire leaf surface, or they can be restricted to certain areas, such as abaxial

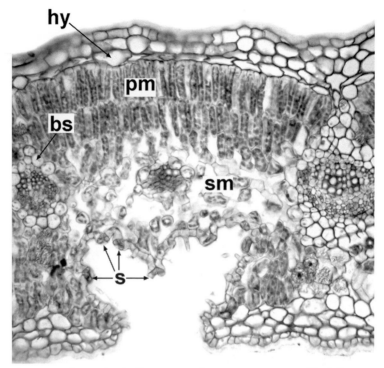

Figure 4.5 *Nerium oleander* (Apocynaceae), transverse section of leaf showing abaxial depression with stomata. bs = bundle sheath, hy = hypodermis, pm = palisade mesophyll, sm = spongy mesophyll. Scale = 500 μm.

surface grooves or leaf margins. Some species possess several different types of trichome on the same leaf. For example, many species of the mint family Lamiaceae characteristically possess two or more sizes of glandular trichome and either branched or unbranched nonglandular trichomes (Figs 1.10, 4.6). Specialized hair types include stinging hairs (e.g. in *Urtica dioica*: Fig. 1.11), water-absorptive leaf scales (e.g. in *Tillandsia*), and salt-secreting glands (e.g. in *Avicennia*, *Limonium* and *Tamarix*[60]). Salt glands accumulate sodium chloride and so allow plants to colonize highly saline soils.

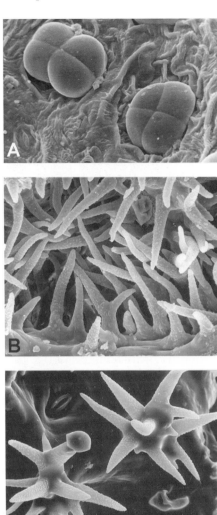

Figure 4.6 Abaxial leaf surfaces of Lamiaceae, showing different trichome types (SEM). (A) Hyptis *caespitosa*, sunken glandular trichomes. (B) Hyptis *proteoides*, surface depression lined with short non-glandular trichomes, mostly uniseriate. (C) Hyptis *emoryi*, two highly-branched trichomes, one with a small gland at the end of one branch.

Papillae are short epidermal projections that are present on the leaf surfaces of some species. For example, within the *Iris* family (Iridaceae), papillae are frequently present on leaf epidermal cells, in some species as a single papilla per cell, and in other species in a row of up to nine papillae per cell[91]. Some species possess a ring of four or more (sometimes coalesced) papillae around each stoma, together forming a raised rim.

4.3.4 Cuticle and Wax

The cuticle is a non-cellular layer composed of several inert polymers, especially cutin, that covers the entire leaf surface and most other aerial plant surfaces[60]. In mesomorphic leaves the cuticle is typically thin and almost transparent, but many xeromorphic plants possess a thick leaf cuticle that often appears lamellated in transverse section. In some species the outer surface of the cuticle possesses characteristic patterns of ridges, folds or striations. These can be short or long and oriented randomly or in a regular pattern, sometimes radiating around stomata or trichomes[9]. Surface patterning may have biological significance in relation to mechanical and optical properties and wettability of the surface. For example, the cuticular patterns on the mature leaf surfaces of *Aloe* and related genera are under precise genetic control[6].

Leaves of some species have a covering of wax over the cuticle. Epicuticular wax is either in the form of a surface crust, or more commonly in small particles of varying shapes and sizes, ranging from flakes to filaments and granules[10]. Wax particles are variously orientated and sometimes occur in characteristic patterns. Certain compounds, such as terpenes and flavonoids, can be recognized by their detailed structure.

4.4 Extrafloral Nectaries

Apart from various types of glandular trichomes, some plants possess specialized nectar-secreting regions (extrafloral nectaries)

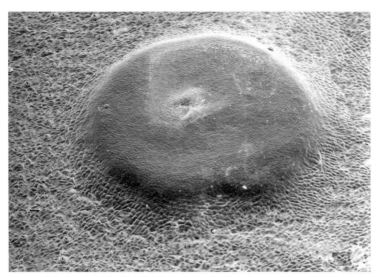

Figure 4.7 *Omphalea diandra* (Euphorbiaceae). Extrafloral nectary on abaxial leaf surface (SEM).

on the leaf or petiole (Fig. 4.7). Although in some cases the function of the extrafloral nectaries remains in doubt, and may be related to regulation of surplus sugars, most are believed to have a role in insect—plant relations. For example, some extrafloral nectaries attract ants (e.g. in *Acacia*), which protect the plant against potential insect herbivores. Extrafloral nectaries are often found over veins, or in the angles of principal veins, or at the proximal or distal ends of the petiole. As with floral nectaries (chapter 5.10), they consist of groups of glandular trichomes (e.g. in some *Hibiscus* species) or regions of anticlinally elongated secretory epidermal cells, often associated with underlying vascular tissue. Extrafloral nectaries are sometimes situated in specialized pockets (domatia), or alternatively occur in pits or raised regions (e.g. in many Euphorbiaceae). Pearl glands, or pearl bodies, occur on leaves of some tropical eudicots (e.g. Fabaceae) and magnoliids (e.g. Piperaceae); these are globular trichomes that are specialized to attract ants, which then protect the plant from herbivory. They secrete substances rich in carbohydrates, lipids and proteins[76].

4.5 Mesophyll

Chlorophyll is contained in chloroplasts in the mesophyll, which is the primary photosynthetic tissue of the leaf. In many plant species the mesophyll is divided into distinct regions (termed palisade and spongy tissues), though in others it is relatively undifferentiated and homogeneous throughout the leaf. Palisade mesophyll is typically adaxial. Palisade cells are anticlinally elongated and possess relatively few intercellular air spaces. By contrast, spongy mesophyll is typically located on the abaxial side of the leaf, and consists of variously-shaped cells with many air spaces between them. Both palisade and spongy tissues can be up to several cell layers thick. Occasionally there is intergradation between the two tissues.

Many tropical grasses, and also some other taxa that are only distantly related (both monocots and dicots), possess a ring of mesophyll cells radiating from the vascular bundles (Kranz anatomy; chapter 4.8). This structure is commonly associated with the C_4 pathway of photosynthesis (Fig. 4.11). In thick leaves, particularly those of some monocots, the central cells are often large, undifferentiated and non-photosynthetic. In the thick "keel" or "midrib" of *Crocus* leaves, a region of large parenchymatous cells with their walls often broken down to form a cavity causes the characteristic white stripe along the centre of the leaf (Fig. 4.8)[91].

In some xeromorphic plants (e.g. *Ilex* and *Ficus*: Fig. 4.4) subdermal layers immediately within the adaxial (or, more rarely, the abaxial) epidermis are modified into a hypodermis. This consists of one or more layers of non-photosynthetic cells that are usually slightly larger and thicker-walled than the adjacent

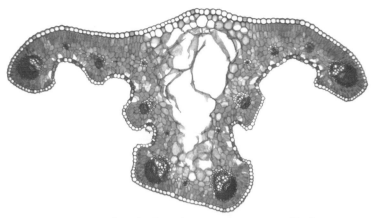

Figure 4.8 *Crocus cancellatus* (Iridaceae), transverse section of leaf. Scale = 100 μm.

mesophyll cells and in transverse section frequently resemble epidermal pavement cells. Hypodermal cells become lignified in some older leaves.

4.6 Sclerenchyma and Idioblasts

Mesophyll is often interspersed with sclerenchyma, particularly at leaf margins and extending as girders from the vascular bundles to the epidermis. Fibres are typically found in groups associated with the vascular bundles or leaf margins, but sclereids are normally isolated in the mesophyll. For example, star-shaped astrosclereids occur in leaves of *Nymphaea* and petioles of *Camellia* (Fig. 1.8). Osteosclereids are characteristic of species with centric leaves, such as *Hakea*. In some species sclereids are associated with veinlet endings.

Other types of idioblast may also be interspersed in the mesophyll. For example, secretory myrosin cells are often found in the leaves of many Brassicaceae, and laticifers occur in leaves of angiosperms, especially those of some eudicots such as *Euphorbia* and *Ficus* (chapter 1.4).

4.7 Leaf Vasculature

There are two main leaf venation types among the angiosperms: parallel and reticulate. Broadly, parallel venation is typical of monocots and reticulate venation of eudicots and magnoliids, though there are many exceptions. In leaves with parallel venation the main veins (primary veins) are parallel for most of their length and converge or fuse at the leaf tip. Typically, numerous small veins interconnect the larger veins, but there are very few vein endings

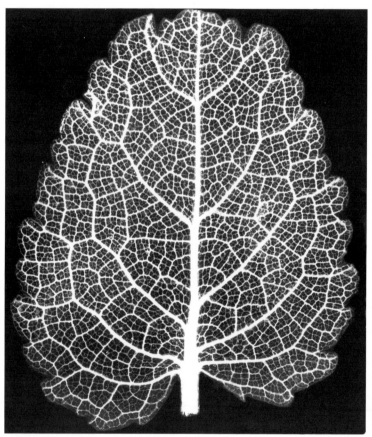

Figure 4.9 Leaf venation: *Hyptis pauliana* (Lamiaceae), cleared leaf with reticulate venation.

in the mesophyll. In leaves with reticulate venation (Fig. 4.9) there is often a major vein in the middle of the leaf, the midrib or primary vein, which is continuous with the major venation of the petiole. The midrib is linked to many smaller secondary (second-order) veins that branch from it and often extend to the leaf margins. Secondary veins sometimes terminate in a hydathode at the leaf margin. In their turn, smaller veins branch from the second- and subsequent-order venation, forming a reticulate network. The areas of mesophyll between the smallest veins in the leaf are termed areoles. In many species small veins branch into the areoles to form vein endings. Variable aspects of leaf venation include the relative number of veinlet endings per areole, and whether second-order veins terminate at the margins or loop around to link with the superadjacent secondary veins[51].

Petioles also possess characteristic venation. The simplest form of petiole vasculature appears in transverse section as a crescent,

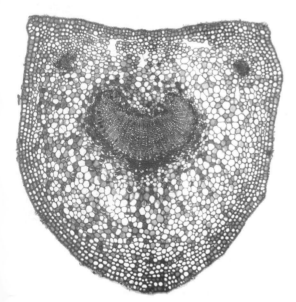

Figure 4.10 *Prunus lusitanica* (Rosaceae), transverse section of petiole. Scale = 100 μm.

with xylem on the adaxial side and phloem on the abaxial side (Fig. 4.10). Some species possess additional bundles outside the main vascular crescent, which may itself be inrolled at the ends, or in a ring, or divided into separate bundles. Classification of the various forms of petiole vasculature depends on how it is linked to the stem vasculature at the node. One or more vascular traces may depart from each gap in the stem vascular cylinder (chapter 2.3). The number and pattern of vascular bundles sometimes vary along the length of the petiole. Midrib vasculature, which is continuous with that of the petiole, is subject to similar variation.

In transverse sections of the lamina, vascular bundles are usually arranged in a single row. However, in some species with very thick leaves, such as *Agave*, there are two or more rows of vascular bundles. Lamina bundles are usually collateral, with adaxial xylem and abaxial phloem, but orientation can vary, and in some cases bundles are bicollateral or even amphivasal. In the isobilateral leaves of some monocots there are two rows of vascular bundles with opposite orientation to each other, the xylem poles being oriented towards the leaf centre. Centric leaves possess a ring of vascular bundles.

Leaf vasculature develops acropetally from the primordial procambial strand at its base[67]. The central trace develops first, and ultimately becomes the midvein. The xylem conducting system of the leaf blade often consists entirely of tracheids, usually with helical or annular thickenings, though in some leaves both vessel elements and xylem parenchyma are also present. The smallest vascular bundles often consist of only one or two rows of xylem tracheids and a few files of phloem sieve tube elements.

4.8 Bundle Sheath and Kranz Anatomy

Most minor vascular bundles in angiosperm leaves are surrounded by a bundle sheath which extends even to the very smallest veins

(Figs 1.13, 4.2). The developmental origin of bundle sheaths differs; some (mestome sheaths) are derived from procambium or vascular meristematic tissue; others are derived from the ground meristem. The bundle sheath typically consists of thin-walled parenchymatous cells, often in a single layer. Some monocots possess distinct inner and outer bundle sheaths, the outer one being parenchymatous and the (partial) inner sheath sclerenchymatous, forming a sclerenchyma cap that is usually located at the phloem pole. In *Aloe* the outer bundle sheath is a specialized tissue that is the source of aloin[12].

Grasses possess either a single sheath consisting of an outer layer of thin-walled cells containing chloroplasts, or a double sheath consisting of an outer layer of thin-walled cells and an inner layer of thicker-walled cells. This is an important taxonomic character in Poaceae, as double sheaths often occur in festucoid grasses and single sheaths in panicoid grasses, though there are exceptions. Leaves of many plants possess regions of

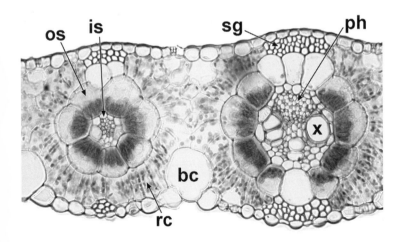

Figure 4.11 *Eleusine coracana* (Poaceae), transverse section of leaf with Kranz anatomy. bc = bulliform cell, is = inner bundle sheath, os = outer bundle sheath, ph = phloem, rc = radiate chlorenchyma, sg = sclerenchyma girder, x = xylem. Scale = 100 μm.

sclerenchyma or parenchyma that extend from the vascular bundle sheaths towards either or both epidermises. These bundle sheath extensions, which are termed girders if they reach the epidermis, afford mechanical support to the leaf and are a xeromorphic feature (Fig. 4.1).

Most plants from warm temperate areas that possess the C_4 pathway of photosynthesis display a modified leaf anatomy that is termed Kranz anatomy (Fig. 4.11). Kranz anatomy occurs in a few groups of both dicots and monocots[97]. It is characterized by elongated mesophyll cells that radiate from a single layer of large parenchymatous bundle-sheath cells containing starch and enlarged chloroplasts. This normally forms a second bundle sheath layer, though in some grasses the primary vascular bundle sheath is itself recruited for this purpose. C_4 plants concentrate CO_2 by photosynthetic carbon assimilation in the radiating mesophyll cells, and subsequent photosynthetic carbon reduction in the bundle sheath cells.

Flower

5.1 Floral Organs

Flowers are complex structures that consist of several organ types borne on a central axis (the receptacle). In many species each flower is subtended by a modified leaf-like structure termed a bract (Fig. 1.2), though bracts are absent from some other species. Within each flower, the organs are arranged in distinct bands (whorls) or in a spiral pattern (Figs 5.1, 5.2). The degree of fusion of individual floral organs within each flower is normally characteristic of a species (i.e. genetically determined). Fusion between similar organ types borne in the same whorl is termed connation. Fusion between different organ types borne in adjacent whorls is termed adnation.

The outer two types of floral organs (collectively the perianth) are modified leaf-like structures, termed sepals (collectively the calyx, or sometimes the first whorl) and petals (collectively the corolla, or the second whorl). In many monocots and magnoliids the perianth organs are morphologically indistinguishable from each other, and are collectively termed tepals, rather than differentiated into sepals and petals. Enclosed within the perianth are the stamens, which are collectively termed the androecium, or sometimes the third whorl, though they often borne in two or more distinct whorls. The carpels (collectively the gynoecium, sometimes termed the fourth whorl) are borne in the centre of the flower, and normally terminate the floral axis.

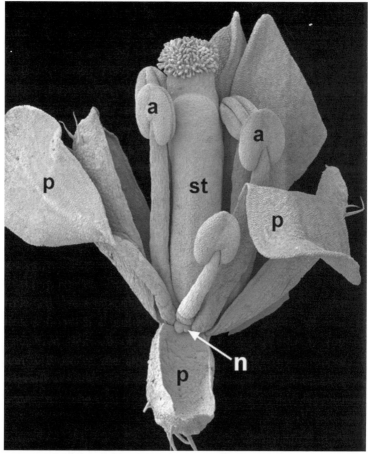

Figure 5.1 *Arabidopsis thaliana* (Brassicaceae), SEM dissected flower. a = anther, n = nectary, p = petal, st = style. Scale = 100 μm.

This general pattern varies considerably among angiosperm groups, and many species possess unisexual flowers. The grass flower is typically subtended by a bract-like structure, the palea, that surrounds three (or, more commonly, two) reduced structures termed lodicules, which are normally interpreted as homologous with a single perianth whorl (probably the inner tepals) of other monocots[4,22,95].

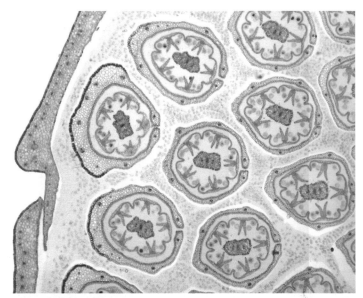

Figure 5.2 *Taraxacum officinale* (Asteraceae), transverse section of inflorescence, showing individual florets. Scale = 100 μm.

Floral organ primordia are typically initiated from the outside inwards, in centripetal (acropetal) sequence towards the floral apex (Fig. 5.3). For example, in *Drimys* the innermost stamens are the last to be initiated in the developing bud, though they are the largest and the first to dehisce in the open flower[113]. However, in some species certain floral organs, either within a single whorl or in more than one whorl, are initiated in a different sequence or in groups (fascicles) (Fig. 5.4). For example, the stamens of some polyandrous palms are initiated centrifugally[115].

5.2 Floral Vasculature

In the majority of flowers, each organ is served by a single vascular strand that diverges from the central vascular cylinder in the receptacle and subsequently branches (Figs 5.5, 5.6). Perianth traces are normally highly branched, often forming a

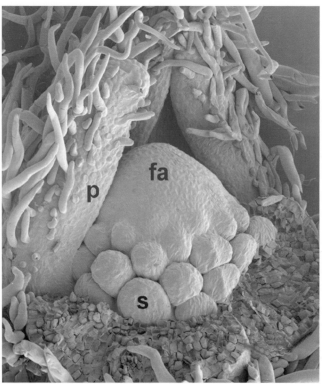

Figure 5.3 *Clematis argentilucida* (Ranunculaceae), SEM dissected flower bud. fa = floral apex, p = developing petal, s = stamen primordium. Scale = 100 μm.

vascular network. Most stamens typically bear a single vascular strand, but some families characteristically possess three or four stamen traces. Others, such as Araceae[35], possess diverse and often branching stamen vasculature. Many species possess two carpellary vascular bundles: the ventral carpellary trace, which diverges into the ovules, and the dorsal carpellary trace, which passes up the style into the stigma (Fig. 5.7). The number of vascular bundles in the style of a syncarpous gynoecium is often an indicator of the number of carpels present, though in some species bundles are branched or fused.

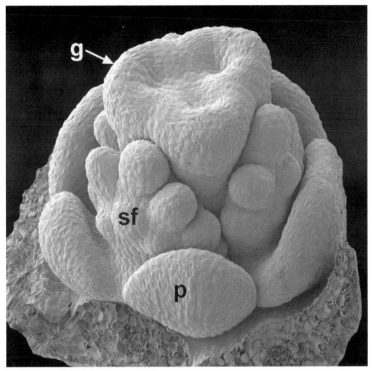

Figure 5.4 *Hypericum empetrifolium* (Clusiaceae), SEM dissected developing flower bud. g = gynoecium, p = petal, sf = stamen fascicle. Scale = 100 μm.

5.3 Perianth

In their simplest form, perianth parts are essentially leaf-like in morphology, though there are many modifications. Sepals are typically green and photosynthetic; it is common to find stomata and trichomes on sepal surfaces. In insect-pollinated plants the petals are typically the largest and showiest part of the flower, though in wind-pollinated plants the petals are sometimes reduced or even absent. Flower colour is largely controlled by the chemistry of the diverse pigments present. Anthocyanins, beta-lains and ultraviolet-absorbing flavonoids are largely confined to the epidermal cells of the petal, whereas other pigments, such

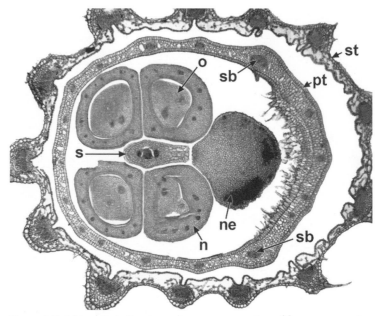

Figure 5.5 *Salvia pratensis* (Lamiaceae), transverse section of flower. n = nutlet, ne = nectary, ov = ovule, pt = petal tube, s = style, sb = stamen vascular bundle, st = sepal tube. Scale = 100 μm.

as carotenoids, occur in either the epidermis or the mesophyll[64]. Petal surfaces frequently lack stomata and the epidermal cells are often domed or papillate (Fig. 5.8), with papillae of various heights and either one or several per cell. The effect of the domed cell surface is to guide incident light into the petal, where it is reflected outwards from the inside walls of the epidermal cells or from the multi-faceted walls of mesophyll cells, thus passing through the pigments in solution in the cell vacuoles. Many petal surfaces are also strongly striated, which may have the effect of further scattering the incident light into the interior of the petal. By contrast, some petal surfaces are smooth (e.g. in *Crocus*), in which case incident light is reflected strongly. In some species of *Ranunculus* with bright yellow flowers, incident light is reflected from starch grains in the subepidermal mesophyll cells[17,78].

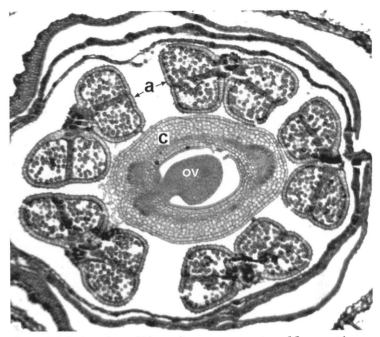

Figure 5.6 *Lupinus arboreus* (Fabaceae), transverse section of flower with five stamens and a single carpel. a = anther, c = carpel, ov = ovule. Scale = 100 µm.

Both petals and sepals consist of an abaxial and adaxial epidermis enclosing usually three or four (or more) layers of undifferentiated isodiametric or elongated cells separated by many air spaces. This mesophyll tissue is interspersed with a row of vascular bundles. As in leaves, some petals and sepals contain idioblasts such as crystal cells, or specialized tissues such as a hypodermis.

5.4 Androecium

The stamens are the pollen-bearing (male) organs. They typically consist of a stalk (the filament), and the pollen-producing part, the anther, which consists of four microsporangia separated into two

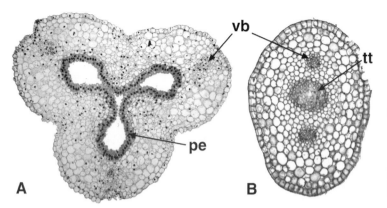

Figure 5.7 Transverse sections of styles: (A) Lilium sp. (Liliaceae), open (hollow) style; (B) *Salvia pratensis* (Lamiaceae), closed style. pe = papillate epidermis, tt = transmitting tissue, vb = vascular bundle (dorsal carpellary trace). Scales = 100 μm.

pairs (thecae), linked by a connective. Each theca possesses two sporangia or anther locules divided by a septum.

Stamen filaments are typically slender and cylindrical, but in some species they are flattened and leaf-like (e.g. in *Nymphaea odorata*) or even branched (e.g. in *Ricinus communis*). In many polyandrous angiosperms the stamens are borne in groups (fascicles). As in other floral parts, the filament surface often bears trichomes, stomata and surface patterning. In transverse section the filament possesses a parenchymatous ground tissue surrounding the vascular tissue, which normally consists of a single vascular bundle.

The anther wall consists of several layers of cells. The epidermis normally undergoes only anticlinal divisions during development. The other anther wall layers are all derived from the primary parietal cells, which arise from the same initial cells as the primary sporogenous cells. The two most distinct anther wall layers are the endothecium, which lies immediately within the epidermis, and the tapetum, which is the innermost layer of cells surrounding the

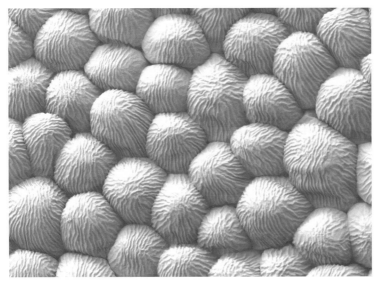

Figure 5.8 *Arabidopsis thaliana* (Brassicaceae), SEM petal surface. Scale = 100 μm.

anther locule (Fig. 5.9). Intervening layers usually consist of thin-walled cells that are often crushed and destroyed at anthesis. Endothecial cells typically develop fibrous wall thickenings which contribute to the anther dehiscence mechanism.

The tapetum is a specialized cell layer that functions as a source of nutrients for developing pollen grains. Tapetal cells are secretory and contain dense cytoplasm. They produce exine precursors, proteins and lipids that form the pollen coat. In many species a layer of tapetal cells remains intact around the anther locule; this type of tapetum is termed the secretory type (or cellular, glandular or parietal). In some other species the tapetal cells degenerate and their protoplasts fuse to form a multinucleate tapetal plasmodium (a periplasmodium) in the anther locule; this type of tapetum is termed the plasmodial type (or amoeboid, invasive, periplasmodial or syncytial type). Transitional types also occur in some species, especially in magnoliids[38].

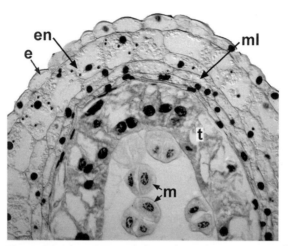

Figure 5.9 *Lilium martagon* (Liliaceae), transverse section of anther wall.
e = epidermis, en = endothecium, ml = middle layer, m = microspores,
t = tapetum. Scale = 100 μm.

5.5 Pollen

In the developing anther the primary sporogenous cells, which
are derived from the same initials as the primary parietal cells,
give rise either directly or by successive mitotic divisions to
the microspore parent cells (microsporocytes). These in turn
each undergo two meiotic divisions (microsporogenesis), either
successively or almost simultaneously, to form a tetrad of haploid
microspores. In the successive microsporogenesis type, callose
cell walls are formed after both meiosis I and meiosis II. In the
simultaneous type, cytokinesis does not occur until both meiotic
nuclear divisions are complete. Microsporogenesis is simultaneous
in most eudicots, whereas both successive and simultaneous types
occur in monocots and magnoliids[39,41].

Following meiosis, the tetrad normally fragments into individual microspores, though in some families (e.g. Winteraceae) the
microspores typically remain together as permanent tetrads. Prior
to anthesis, each microspore undergoes an unequal (asymmetric)

mitosis to form a larger vegetative cell and a smaller generative cell enclosed within the pollen grain wall. The generative (spermatogenous) nucleus later undergoes a further mitosis (spermatogenesis) to form two sperm cells, either in the pollen grain or in the pollen tube. The microgametophyte is therefore either bicellular or tricellular. Spermatogenesis is itself asymmetric in some angiosperms (e.g. *Plumbago* and *Zea*), in which dimorphic sperm are produced that can preferentially fertilize either the egg nucleus or the polar nuclei in the embryo sac[114].

Pollen grains are radially or bilaterally symmetrical bodies that represent units of dispersal from the anther to the stigma (Fig. 5.10). They vary considerably in size and shape among species[29]. The pollen-grain wall consists of two distinct domains: a hard outer exine, which is composed mainly of sporopollenin (a carotenoid polymer), and a relatively soft inner intine composed of polysaccharides. The exine is itself a layered structure, often differentiated into an outer sculptured ectexine (sexine) and an inner non-sculptured endexine (nexine), though endexine is normally well-developed only in eudicots with tricolpate pollen[41].

Apertures are often present in the pollen-grain wall, though pollen grains of some species lack a clearly defined aperture (termed inaperturate pollen[37]). Apertures represent specialized regions in which the outer layer (the exine) is reduced or absent, and the underlying layer (the intine) is thickened. They range in shape from elongated furrows to roughly circular pores. Some apertures possess a lid-like operculum[40]. Apertures that lie along the distal face of the pollen grain (normally the face that was directed outwards in the tetrad) are termed sulci. Sulcate pollen grains are characteristic of monocots and magnoliids. Apertures that lie along the equatorial face of the pollen grain, as defined during the tetrad phase, are termed colpi. Pollen with three equatorial apertures (tricolpate pollen) characterizes the eudicots (sometimes termed the tricolpates). In many eudicots, microspore

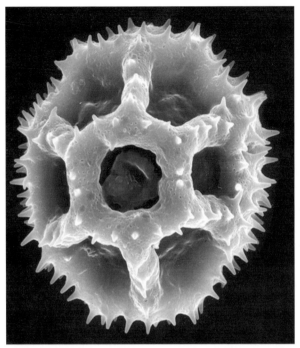

Figure 5.10 *Taraxacum officinale* (Asteraceae), SEM pollen grain. Scale = 10 μm.

tetrads are tetrahedral with three apertures (colpi) arranged equidistantly around the equator of the microspore. Colpi are typically elongated and slit-like, but can occasionally be reduced to pores, and in some eudicot species the number can increase to four, five, six, or more. In spiraperturate pollen grains (e.g. some *Crocus* species) the aperture spirals around the grain.

Pollen grains dehydrate after contact with the air, and the exine contracts. Subsequently, rehydration and exine expansion occurs on the stigmatic surface. Dry and hydrated pollen grains of the same species can appear very different in size, shape and even surface features. There are many different patterns of exine sculpturing. For example, the exine may be reticulate or areolate, or it may possess surface holes (puncta), granules, warts or spines[84].

These differences are often of considerable taxonomic significance. The surface patterning is a mechanical adaptation, either ensuring elasticity of the wall or helping to accommodate changes in pollen shape associated with hydration and subsequent increase in cytoplasmic volume (harmemogathy). Wind-borne pollen grains are generally small and light, and possess relatively little surface sculpturing. Water-dispersed pollen often possesses adaptations such as a slime coating (e.g. in Hydrocharitaceae[65]).

In some species with animal-dispersed pollen, substances such as lipids, proteins and carbohydrates are stored and dispersed with the pollen in the intercolumellar spaces of a deeply-chambered exine. The substances are normally derived from the tapetum. They have various functions, such as conferring odour, or causing grains to aggregate into sticky masses, which is useful for effective animal dispersal. In some species (e.g. many Brassicaceae, Malvaceae and Asteraceae[48]), exine-borne substances are released onto the stigma when the exine expands after pollen grain rehydration, and hence play a role in the control of interspecific compatibility.

5.6 Gynoecium

The carpels are the ovule-bearing (female) organs. Collectively they form the gynoecium, which consists of an ovary, style(s) and stigma(s). In syncarpous gynoecia the carpels are fused, either postgenitally, if they are initiated separately and become fused during development, or congenitally, if the gynoecium is initiated as a single structure.

5.6.1 Stigma and Style

Stigmatic epidermal cells provide a receptive surface for pollen grains (Fig. 5.11). They are typically secretory, with a specialized cuticle, and may be slightly domed or possess variously elongated papillae. Stigmas of some species possess little or no surface secretions (termed dry stigmas), whereas "wet" stigmas exude copious surface secretions that play a role in pollen recognition and germination[49]. The stigmatic cuticle is often stratified in transverse section, with a lamellated outer layer and reticulate inner layers. Some *Crocus* species possess a chambered cuticle, and in some *Euphorbia* species the cuticle is fenestrated[47].

Rarely, the stigma is borne directly on the top of the ovary, but most commonly it is borne on a style. The ground tissue of the style is parenchymatous, and it typically possesses two or

Figure 5.11 *Arabidopsis thaliana* (Brassicaceae), SEM stigma. Scale = 100 μm.

three dorsal carpellary traces, depending on the number of carpels. Many syncarpous eudicots possess "solid" styles, in which there is a central specialized secretory tissue, the transmitting tissue, that links the stigma with the centre of the ovary, and serves as a nutrient-rich tract for pollen-tube growth[110] (Fig. 5.7). The transmitting tissue is derived from superficial carpellary tissue of the fused carpel margins. By contrast, most syncarpous monocots and some eudicots (e.g. *Acer saccharum*[81]) possess "open" styles with a central stylar canal that is often filled with mucilage (Fig. 5.7), though grasses possess solid styles with a dense central trans-mitting tissue.

5.6.2 Ovary

In many syncarpous species there is an opening (compitum) at the base of the style that allows the pollen tubes to reach any of the ovules in the ovary locules. The ovary contains one or more ovules (Fig. 5.12). Each ovule is attached to the ovary wall at a placenta. The arrangement of the placentae and the numbers of carpels and ovary locules vary in different species. In many syncarpous species the numbers of ovary locules and carpels are equivalent, but this can vary; for example in most orchids the ovary is syncarpous, tricarpellary and unilocular.

Ovules are borne on placentas, which are meristematic regions borne at the carpel margins within the ovary locules[100]. The arrangement of the placentas (placentation) varies among species. In species with two or more locules in which the ovules are borne on placentas in the central axis, placentation is axile. Parietal placentation occurs in some species with unilocular ovaries, though false septa occur in some species. Other placent-ation types occur in unilocular ovaries; for example, the placentas can be at the base of the ovary (basal placentation), or on top of a central column of tissue that is not joined to the ovary wall except at top and bottom (free central placentation).

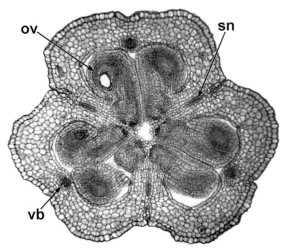

Figure 5.12 *Asparagus officinale* (Asparagaceae), transverse section of ovary. ov = ovule, sn = septal nectary, vb = vascular bundle. Scale = 100 μm.

Many species possess proliferations of secretory tissue around the bases of the funicles, providing nutrients for the developing pollen tubes and guiding them into the micropyles. These structures, termed obturators, are derived either from the placenta or the funicle, and are frequently papillate[110].

5.7 Ovule

Ovule primordia are initiated as small swellings in the placenta. The ovule consists of a nucellus, which bears the embryo sac, enclosed by one or two (inner and outer) integuments (Figs 5.13–5.15). The region where the nucellus, integuments and funicle unite is termed the chalaza (Fig. 6.1). The micropyle is a narrow opening in the ovule formed by one or both integuments, located at the opposite end of the embryo sac to the chalaza. At anthesis, each ovule is attached to the ovary by a funicle, which normally possesses a single vascular strand. The vascular bundle in the funicle usually terminates at the chalaza, but in some species it is more extensive.

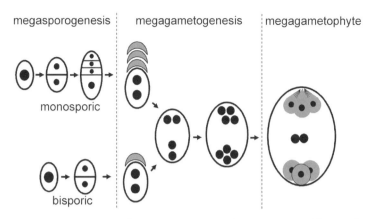

Figure 5.13 Diagram of megagametogenesis (monosporic and bisporic) and megasporogenesis.

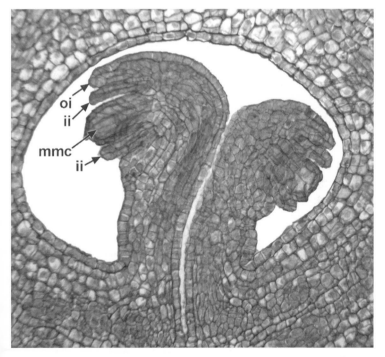

Figure 5.14 *Bomarea mexicana* (Alstroemeriaceae), transverse section of ovary showing ovule primordia. ii = inner integument, mmc = megaspore mother cell, oi = outer integument. Scale = 100 μm.

The nucellus arises from the apex and body of the ovule primordium. The integuments develop from around the primordium base and encircle its apex, forming the micropyle; normally the inner integument is initiated before the outer. The possession of two integuments is the most common condition in angiosperms, but a single integument characterizes some eudicots, and a few species lack integuments entirely[16]. Prior to formation of the archespore, the ovule primordium is organized into either two or (more commonly) three distinct zones defined by orientation of cell division. The outermost (dermal) layer initially undergoes mainly anticlinal divisions, but may later proliferate at the micropylar end to form a nucellar cap, and sometimes also proliferates around and below the embryo sac. The archespore forms in the subdermal layer, immediately subtended by the central zone.

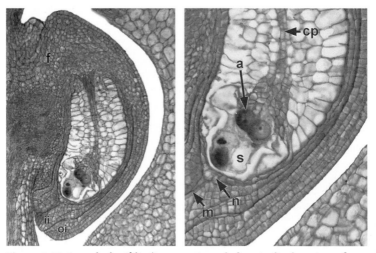

Figure 5.15 Lomandra longifolia (Laxmanniaceae), longitudinal section of ovule (left) and embryo sac (right). a = antipodal cell, cp = conducting passage, ii = inner integument, m = micropyle, n = nucellus, oi = outer integument, s = synergid. Scales = 100 μm.

In some species the nucellus proliferates in the central zone to form a hypostase, which is a chalazal region of often refractive and suberinized or lignified tissue. For example, a large hypostase is present in *Acorus* and *Crocus*. Sometimes other specialized structures are formed in the central zone of the nucellus, such as a postament or conducting passage (Fig. 5.15); these often persist after fertilization. The dermal nucellar cells surrounding the embryo sac often break down before (or shortly after) fertilization, in which case the innermost epidermal layer of the inner integument is sometimes differentiated to form an endothelium or integumentary tapetum (Fig. 6.6). Endothelium cells are frequently enlarged and densely cytoplasmic, and sometimes endopolyploid; in some species they have a secretory function.

5.8 Embryo Sac

Within the nucellus, a single (normally hypodermal) cell becomes a primary sporogenous cell (archesporial cell, or archespore). The archespore rarely consists of more than one cell, though it can be multicellular in a few species (e.g. *Brassica campestris*[86]), in which one cell produces the megagametophyte. In turn, the archesporial cell either gives rise directly to the megaspore mother cell (megasporocyte) or undergoes mitosis to form a primary parietal cell and a megasporocyte. The megasporocyte then undergoes two meiotic divisions (megasporogenesis) to form a tetrad of four megaspores, which are usually either in a linear or T-shaped arrangement. In the majority of angiosperm flowers, one megaspore (most commonly the chalazal one) gives rise to the mature embryo sac by further mitotic divisions, and the other three megaspores degenerate (Fig. 5.13). This type of development is termed monosporic. However, in relatively few angiosperms, two or four megaspores play a role in embryo sac formation; these types are termed bisporic or tetrasporic respectively[69,125,126]. Degenerated megaspores are often surrounded by persistent callose[68].

In most angiosperms the mature embryo sac (megagametophyte) possesses eight nuclei arranged in seven cells (Fig. 5.15), though types with four and sixteen or more nuclei have also been recorded. The most common type is monosporic and eight-nucleate; this is sometimes termed the *Polygonum* type of embryo sac development. At the binucleate stage, the two nuclei migrate to the micropylar and chalazal poles and subsequently divide. Of the two micropylar nuclei, the one closest to the micropyle divides to form the synergids, and the other divides to form the egg cell and one of the polar nuclei. The two chalazal nuclei each divide so that one forms two antipodal cells and the other forms an antipodal and a polar nucleus. The two polar nuclei migrate to the centre and fuse to form a diploid fusion nucleus. Cellularization follows, so that the mature megagametophyte consists of three antipodal cells at the chalazal end, a central cell with a fusion nucleus, and two synergids plus an egg cell at the micropylar end.

The synergids and the egg cell are so tightly pressed together that they are collectively termed the egg apparatus. The synergids play a role in directing the pollen tube into the embryo sac; they are calcium-rich and normally possess a series of wall thickenings, the filiform apparatus, which extends into the micropyle (Fig. 5.15). In many species the antipodals degenerate at an early stage, but in others they persist, and sometimes undergo cell division (e.g. in many grasses) or endoreduplication.

5.9 Pollen-Tube Growth

Pollen tubes emerge from a pollen grain that has landed on a conspecific receptive stigma, normally via an aperture in the exine. The pollen-tube wall is initially formed from the innermost

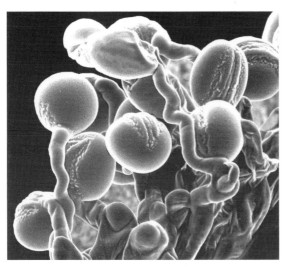

Figure 5.16 *Anomatheca laxa* (Iridaceae), SEM germinating pollen grains on stigmatic surface.

(cellulosic-callosic) layer of the intine; in some species this is then locally dissolved and replaced by an emergent outgrowth of the plasma membrane of the vegetative cell[50,79,101]. The pollen tube grows at the apex, leaving an attenuated, loosely-arranged microfibrillar wall interspersed with polysaccharide particles. Germinating pollen tubes (Fig. 5.16) grow towards the ovules either intrusively between cells of transmitting tissue (in solid styles) or over a layer of mucilage (in hollow styles). They obtain nourishment (polysaccharides and proteins) from the richly cytoplasmic transmitting tissue (in solid styles) or the glandular epidermis (in hollow styles), and subsequently from specialized secretory tissues such as the obturator or the outer integument of the ovule.

Pollen tubes, though typically unbranched, can become branched in some species (e.g. *Grevillea*), in which one (siphonogamous) branch enters the micropyle and other (presumably haustorial) branches invade adjacent ovary or ovule tissue[59].

Further remarkable examples of the intrusive abilities of pollen tubes are provided by rare cases in which pollen germinates prior to anther dehiscence and pollen tubes grow through the anther filament tissue and enter the female tissue at the receptacle, thus bypassing the stigma (e.g. in the monocot *Sagittaria*, the eudicot family Malpighiaceae and a gametophytic *Arabidopsis* mutant[1,58,120]). Some angiosperms (e.g. *Lavatera*) exhibit polysiphony: multiple (15–20) pollen tubes emerge from a single region of the grain on the side in contact with the stigma, though only one tube bears the sperm[75].

When the pollen tube reaches an ovule, it enters through the micropyle and discharges the male gametes (sperm cells) before collapsing. One sperm enters the haploid egg cell and the other the diploid central cell, and nuclear fusion ensues, resulting in a diploid zygote and a triploid primary endosperm nucleus. This process, termed double fertilization, is one of the main characteristics of the angiosperms. Subsequently, the zygote and primary endosperm nucleus divide to form the embryo and endosperm respectively (chapter 6).

5.10 Floral Secretory Structures

Many flowers bear specialized secretory structures, such as nectaries, elaiophores and osmophores. These secrete nectar, oil and scent respectively, to attract potential pollinators, including insects such as bees and moths, and also vertebrates such as humming birds and bats.

Nectaries are localized areas of tissue that regularly secrete nectar, a sugar-rich substance that is attractive to animals[31]. Nectaries usually consist of secretory epidermal cells with dense cytoplasm, sometimes modified into trichomes. Adjacent sub-epidermal cells may also be secretory, and in some cases nectar passes to the surface through modified stomatal pores.

Vascular tissue close to the nectary often consists mainly or entirely of phloem, which transports sugars to the secretory region.

Nectaries may occur on any floral organ, or they may represent an entire modified organ, or even a novel structure. Most species of the mint family (Lamiaceae) possess an enlarged nectariferous disc surrounding the base of the ovary (Fig. 5.5), which is derived from developing ovary tissue. In most Brassicaceae (e.g. *Arabidopsis thaliana*) the nectary is located at the base of the stamen filament (Fig. 5.1). Septal nectaries are characteristic of many monocots; these occur at the unfused carpel margins in monocots with postgenitally fused ovaries (Fig. 5.12). Nectar produced from septal nectaries is exuded from secretary epidermal cells and emerges from small pores or slits on the surface of the gynoecium[25, 93].

Flowers of some insect-pollinated species lack nectaries, but possess other secretory structures that attract potential pollinators. For example, some flowers bear oil-secreting glands, termed elaiophores (Fig. 5.17), which are morphologically similar to some types of nectary[118]. Osmophores[119] are modified floral

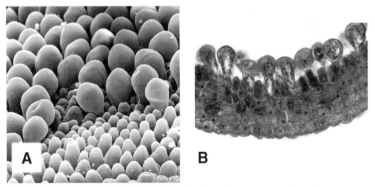

Figure 5.17 *Tigridia meleagris* (Iridaceae), elaiophores on tepal surface (A) SEM, (B) LM. Scale = 100 μm.

structures that produce volatile secretions (scents); in some orchids the odours are pheromone-like substances. Osmophores typically possess a relatively thick, domed or papillate epidermis with densely cytoplasmic contents. In *Platanthera bifolia* the epidermis of the labellum secretes a nocturnal scent, and in *Ophrys* species osmophores on the labellum consist of areas of dome-shaped, papillate, dark-staining epidermal cells. Flowers of *Narcissus* emit pollinator-specific volatiles that are probably derived from the colourful corona[5].

Seed and fruit

6.1 Seed Coat

The seed coat prevents destruction of the seed by dehydration or predation. In bitegmic seeds the testa is derived from the outer integument, and the inner integument forms the tegmen[15] (Fig. 6.1). In unitegmic seeds the term "testa" applies to the entire seed coat. Seed coats are multilayered tissues; they generally

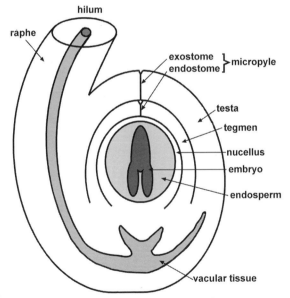

Figure 6.1 Seed organization: Diagram of a generalized campylotropous bitegmic dicotyledonous seed with perisperm. (Adapted from Boesewinkel and Bouman 1984.)

include a hard, protective mechanical layer that is formed from all or part of the testa or tegmen[23]. In exotestal seed coats the mechanical layer is derived from the outer epidermis of the outer integument, whereas in endotegmic seed coats it is derived from the inner epidermis of the inner integument. In some species the mechanical layer consists of one or more rows of elongated, palisade-like cells, such as the macrosclereids in the exotesta of many Fabaceae.

Seed coat surfaces exhibit a variety of cellular patterns, often with characteristic papillate or striate surface sculpturing[9] (Fig. 6.2). Some seeds possess epidermal trichomes; for example, the seed coat hairs of *Gossypium* (cotton) are an important source of textile fibres.

Seed-coat vasculature usually consists of a single bundle passing from the raphe to the chalaza, but this can vary in extent and degree of branching. Many seed coats possess specialized structures that are related to dispersal[15]. For example, some wind-dispersed seeds possess wings, and some animal-dispersed seeds are fleshy.

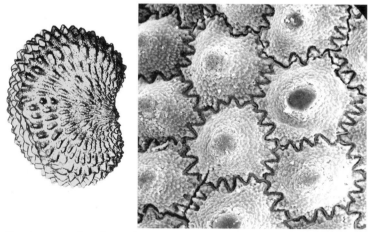

Figure 6.2 Seed surface: *Silene nutans* (Caryophyllaceae), entire seed (left) and detail (right) showing papillate epidermal cells with sinuous anticlinal walls.

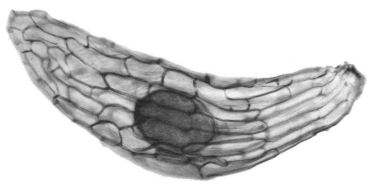

Figure 6.3 Dust seed: *Cypripedium calceolus* (Orchidaceae), light micrograph of entire seed, showing thin testa surrounding globular embryo. Scale = 100 μm.

The fleshy part of the seed coat, termed the sarcotesta, is most commonly formed from part of the outer integument. Arils are fleshy outgrowths of the funicle. Some plants, especially parasitic or mycoheterotrophic plants such as *Monotropa* or orchids, produce large numbers of highly reduced "dust seeds" from each ovary; these minute seeds can be blown over long distances. The seed coat of many orchids lacks vasculature entirely (Fig. 6.3).

6.2 Pericarp

In fruits that are derived from a single ovary, the fruit wall, termed the pericarp, is typically derived from the ovary wall. The pericarp displays a similar range of variation to the seed coat, depending particularly on whether it is dry or fleshy and dehiscent or indehiscent. It is typically divided into three layers – the outer exocarp, central mesocarp and inner endocarp – though in some fruits the three layers are not readily distinguished. At least one layer of the fruit wall often consists of thick-walled lignified cells (Fig. 6.4), though in some fleshy fruits (berries), such as those of *Vitis vinifera* (grape), the entire endocarp consists of thin-walled succulent cells. In other fleshy fruits (drupes), such as those

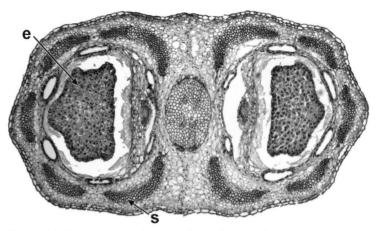

Figure 6.4 Fruit anatomy: *Anthriscus sylvestris* (Apiaceae), transverse section of entire fruit, showing two endospermous seeds. e = endosperm, s = sclerenchyma in pericarp wall. Scale = 100 μm.

of *Prunus persica* (peach), the endocarp cells are thick-walled and only the mesocarp is fleshy, the exocarp being a narrow epidermal layer. In *Olea europaea* (olive) the fleshy mesocarp is interspersed with thick-walled sclereids.

Many indehiscent seeds and fruits produce sticky mucilage when they become wet, which provides an adhesive for animal-mediated dispersal, a phenomenon termed myxospermy or myxocarpy[96]. For example, in the nutlet walls of *Coleus* and some other Lamiaceae the epidermal cells absorb water and then rupture, producing large amounts of slime interspersed with coiled, thread-like protuberances.

6.3 Grass Caryopsis

The grass "seed" is actually an indehiscent one-seeded fruit in which the testa and pericarp are fused together to form a caryopsis (Fig. 6.5). The grass caryopsis is an indehiscent fruit (an achene) in which the seed coat has undergone further reduction[95].

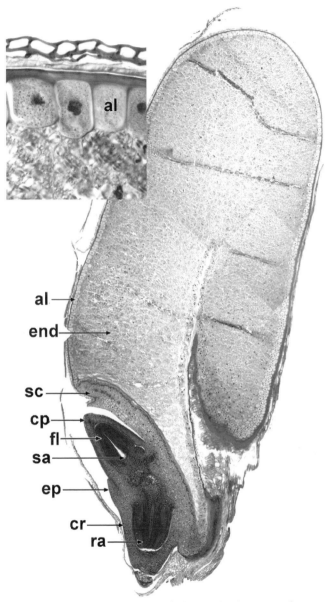

Figure 6.5 *Triticum vulgare* (Poaceae), longitudinal section of caryopsis, with detail of aleurone layer (inset). al = aleurone layer, cp = coleoptile, cr = coleorhiza, end = endosperm, ep = epiblast, fl = foliage leaves, ra = radicle, sa = shoot apex, sc = scutellum. Scale = 1 mm.

After fertilization the pericarp consists of a few cell layers, and the integuments disintegrate completely, leaving only a hyaline membrane covered with a cuticle, derived from the outer layer of the inner integument. Grass seeds also possess highly differentiated embryos with a unique highly characteristic prominent outgrowth of the embryo, termed the scutellum, which is normally considered to be a modified cotyledon. Some grasses possess an outgrowth opposite the scutellum, termed the epiblast, which has been variously interpreted as a second cotyledon or an outgrowth of the first cotyledon or of the coleorhiza. Grass embryos are well-differentiated within the seed, prior to germination. They characteristically possess a sheath (coleoptile) surrounding the epicotyl and plumule, and a well-developed radicle also surrounded by a sheath (the coleorhiza). In some grasses (and a few other angiosperms) the outermost layer of the endosperm, termed the aleurone layer, is a specialized tissue of enlarged cells containing protein bodies and large nuclei.

6.4 Endosperm

Endosperm forms a food-storage tissue in the seed. It not only promotes growth and longevity for the seed, but can also facilitate dispersal, as an attractant to animals. Mature endosperm typically consists of tightly-packed cells that contain food-reserve materials such as starch grains or protein bodies. Endosperm is typically a triploid tissue formed by fusion of one sperm cell with two female polar nuclei. It is present in most angiosperm seeds but in greatly contrasting amounts; for example, endosperm formation is negligible in orchid seeds but extensive in grass seeds, in which it forms an important economic crop.

Early endosperm development is traditionally classified into three types, termed nuclear, cellular and helobial, based on the timing and degree of cell wall formation, though transitional forms exist[117]. Nuclear endosperm is the most common type, and occurs in many eudicots (e.g. *Arabidopsis thaliana*). Nuclear endosperm possesses both a syncytial (free-nucleate) phase and a cellular phase. Early cell divisions are not followed by cell wall formation, and the nuclei are initially free in the cytoplasm of the embryo sac, usually surrounding a central vacuole. Cell walls eventually form in most endosperm tissues, but sometimes the nuclei at the chalazal end remain free; for example, the liquid "milk" of the coconut palm (*Cocos nucifera*) is a syncytium that contains many free endosperm nuclei in addition to oil droplets and protein granules.

In the cellular type of endosperm formation, which occurs in some eudicots (e.g. Acanthaceae), even the earliest nuclear divisions are followed by cell-wall formation. In the helobial endosperm type, which is restricted to some monocots, the primary endosperm nucleus undergoes division to form two unequal chambers, normally a small chalazal chamber and a large micropylar chamber. The nucleus of the micropylar chamber migrates to the top of the embryo sac, and its initial divisions are not accompanied by cell wall formation, though cell walls are formed with later mitoses. The chalazal chamber has far fewer nuclear divisions, and its nuclei remain free in the cytoplasm; it typically has a haustorial role.

Endosperm haustoria may develop in all three types of endosperm. Haustoria assist nutrient absorption and sometimes invade adjacent tissues. For example, most species of the mint family Lamiaceae possess both chalazal and micropylar haustoria (Fig. 6.6), which may be either free-nucleate or cellular, sometimes even amoeboid[89]. In these species, the first division of the primary endosperm nucleus is longitudinal, followed by formation of a transverse wall. The chalazal nucleus forms a small chalazal haustorium close to the antipodals, and the micropylar nucleus divides further to form a micropylar haustorium and a central cellular endosperm. The micropylar haustorium transfers nutrients from the integument to the embryo and cellular endosperm.

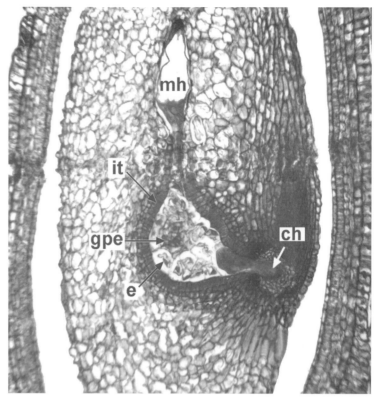

Figure 6.6 *Prunella grandiflora* (Lamiaceae), longitudinal section of fertilized seed showing endosperm with haustoria. ch = chalazal haustorium, e = endosperm, gpe = globular proembryo, it = integumentary tapetum, mh = micropylar haustorium. Scale = 100 µm.

The chalazal haustorium transfers nutrients from the vascular bundle to the endosperm.

6.5 Perisperm

In some plants, especially monocots, parts of the nucellus enlarge or proliferate after fertilization, and have a role as a regulating or storage tissue for the developing embryo. Seed storage tissues derived from the nucellus are termed perisperm[92]. In some monocots (e.g. some members of the order Poales) endosperm is entirely absent from the mature seed, and perisperm represents the primary storage tissue[95]. Seeds of *Yucca* possess perisperm that contains membrane-bound protein and oil bodies within the cells, together with reserve carbohydrates in the thick cell walls[53]. Some members of the monocot order Zingiberales (gingers, bananas and their relatives) possess perisperm, but this is often entirely compressed in the mature seed, with only the cell walls remaining (e.g. *Musa*[44]). In other members of Zingiberales (e.g. *Canna*[45]), mitotic activity during ovule development causes the chalaza region of the nucellus to become massive; this region is then sometimes termed a pachychalaza. In *Acorus* the perisperm is dermal in origin, formed from nucellar epidermal cells that elongate and become filled with transparent proteinaceous cell contents.

6.6 Embryo

In a normal angiosperm reproductive system the embryo develops from the diploid fertilized egg cell (zygote). Following fertilization, the zygote often undergoes a change in volume, either shrinkage or enlargement, before cell division commences. The initial cell division is usually transverse and sometimes asymmetric, to form a small apical and larger basal cell[123]. The pattern of subsequent cell division varies among species, and has been classified into several types[74]. Most embryos eventually differentiate into an undifferentiated globular mass of cells (the proembryo) attached to the embryo sac wall by a stalk (the suspensor) (Fig. 6.7). In *Arabidopsis* the apical cell gives rise to the proembryo, which ultimately forms the bulk of the embryo, and the basal cell produces the suspensor and the hypophysis, which is the precursor to the root cortex initials and the central region of the root cap. The proembryo can be massive (e.g. *Degeneria*), or small, as in *Capsella*, in which it consists of only eight cells[102].

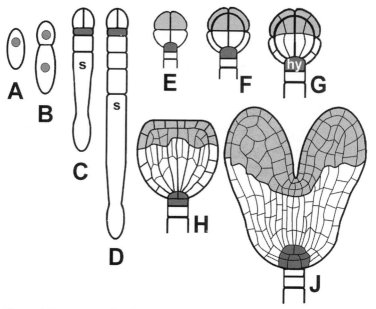

Figure 6.7 Diagram of embryo development. (A) zygote. (B) two-celled stage, result of unequal cell division to smaller apical and larger basal daughter cells. (C, D) growth of suspensor. (E–G) development of globular proembryo. (H, J) development of heart-shaped embryo. hy = hypophysis, s = suspensor.

The suspensor exhibits great diversity in angiosperms[74,127]; it can be uniseriate or multiseriate, and filamentous, spherical or irregular in shape. Cells of large suspensors, such as those of *Phaseolus*, are often endopolyploid. Suspensors of some species are secretory, and those of others (e.g. *Sedum* and *Tropaeolum*) produce haustoria that invade surrounding endosperm tissue.

The suspensor ultimately degenerates, and the globular proembryo undergoes a process of irregular meristematic activity that results in a shift from radial to bilateral symmetry. The proembryo eventually becomes organized into a structure with root and shoot apices at opposite ends of an embryonic axis (the hypocotyl). Embryos of most eudicots and magnoliids

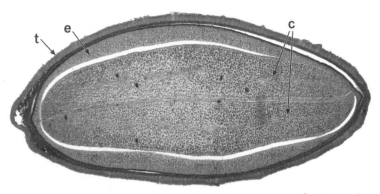

Figure 6.8 *Linum usitatissimum* (Linaceae), transverse section of entire seed. c = cotyledon, e = endosperm, t = testa. Scale = 100 μm.

become bilobed or heart-shaped (Fig. 6.7) as two cotyledons differentiate (Fig. 6.8). Monocot embryos develop a single, often elongated cotyledon. The degree of differentiation of mature embryos varies considerably; for example, in orchids the embryo remains a simple undifferentiated mass of cells (Fig. 6.3). Some highly differentiated embryos possess, in addition to the hypo-cotyl and cotyledons, a short primordial root (radicle), often with a root cap, and a shoot bud or short shoot (epicotyl) developed beyond the cotyledons.

6.7 Seedling

At germination the testa is ruptured and the seedling radicle emerges through the micropyle and pushes through the substrate. Seedlings possess a root (radicle) and a hypocotyl, which bears the cotyledons (seed leaves) and shoot apex. The hypocotyl varies in size and form, from a swollen food-storage organ to a very short structure which may be almost non-existent, in which case the radicle extends almost to the cotyledonary node. Following the emergence of the radicle, either the hypocotyl elongates and the cotyledons and shoot apex emerge (termed epigeal germination), or the cotyledons remain enclosed in the testa and the internode above them (the epicotyl) elongates, pushing the shoot apex upwards (termed hypogeal germination). Epigeal germination is the most common type in angiosperms; the cotyledons are borne above ground, and are usually photosynthetic. By contrast, some larger-seeded eudicots such as legumes (e.g. *Vicia faba*) are hypogeal, and possess fleshy, swollen cotyledons.

In monocots the radicle withers at an early stage, and subsequent roots are shoot-borne (adventitious); they are each initially surrounded by a sheath (coleorhiza), which develops from outer cortical tissue by cell elongation. The cotyledons are usually morphologically different from the first foliage leaves; they typically possess simpler vasculature which often consists of a single vascular bundle. In monocot seedlings the cotyledon typically includes three parts[109]: a basal sheath, a ligule or ligular sheath, and a limb, though the relative differentiation of each region varies; for example, in *Tigridia* seedlings both the hypocotyl and the basal sheath are extremely reduced.

Glossary

abaxial: away from axis (the abaxial leaf epidermis is usually the lower one). cf. adaxial.

abscission layer: well-defined region of tissue separation, e.g. for abscission of leaf from stem.

acropetal: towards the apex (cf. basipetal).

adaxial: towards the axis (the adaxial leaf surface is usually the upper one). cf. abaxial.

adventitious roots: stem or leaf-borne roots.

aerenchyma: specialized parenchymatous tissue normally associated with aquatic plants, with a regular, well-developed system of intercellular air spaces.

aleurone layer: specialized outermost cell layer of endosperm.

amphistomatic leaf: one with stomata present on both surfaces.

amphivasal vascular bundle: one with xylem surrounding phloem.

amyloplast: plastid containing starch.

androecium: collective term for stamens.

anisocytic: one of the classification types of mature stomata; with three unequal subsidiary cells.

anomalous secondary growth: secondary growth that does not fit the "normal" pattern of xylem and phloem production; e.g. xylem with included phloem.

anomocytic: one of the classification types of mature stomata; subsidiary cells absent.

anther: part of stamen bearing pollen (in anther locules).

anthesis: opening (dehiscence) of anther to release pollen; sometimes applied to opening of the flower bud.

anticlinal: perpendicular to the plant surface.

antipodal cell: part of megagametophyte; one of a group of (typically three) cells at chalazal end of mature embryo sac.

aperture (of pollen grain): thin or modified region of pollen exine, through which the pollen tube grows at germination.

apocarpous gynoecium: one in which carpels are not fused.

apoplast: cell areas not bounded by plasmalemma (i.e. cell wall, middle lamella, intercellular spaces).

apotracheal parenchyma (in secondary xylem): axial parenchyma not associated with vessels.

archesporial cell (archespore): primary sporogenous cell (or tissue).

areole: region of mesophyll between smallest veins in leaf.

aril (in seeds): fleshy outgrowth of funicle.

articulated laticifer: one composed of several cells.

astrosclereid: star-shaped or highly branched sclereid.

bark: part of woody stem outside secondary xylem; i.e. including vascular cambium, phloem, cortex and periderm, though sometimes applied only to periderm and outer cortex.

basipetal: towards the base; i.e. away from apex (cf. acropetal).

bast fibre: extraxylary fibre in stem; i.e. cortical or phloem fibre.

bicollateral vascular bundle: one with phloem on both sides of xylem.

bifacial (dorsiventral) leaf: one with both adaxial and abaxial surfaces, usually morphologically different from each other.

bitegmic seed: one with two seed coat layers, derived from two integuments.

brachysclereid (stone cell): more or less isodiametric sclereid.

bulliform cells (usually in grass leaves): groups of epidermal cells that are markedly larger than neighbouring epidermal cells.

bundle sheath: layer of cells surrounding leaf vascular bundles.

callose ($\beta$$-$1,3 glucan): product of plasma membrane that is primarily a component of the cell wall; acts as a permeability barrier or sealant in developing tissues such as microspore tetrads or pollen tubes, or in response to wounding or pathogens.

callus tissue: undifferentiated mass of thin-walled cells; usually wound tissue.

cambium: meristematic band of cells; e.g. cork cambium or vascular cambium.

Casparian strip (Casparian thickening): band of suberin deposited in primary cell walls of root endodermis.

cellulose: primary carbohydrate component of plant cell walls; a large linear polymer composed of D-glucose units with $\beta$$-$1,4 linkages.

centric (terete) leaf: one that is cylindrical, or circular in transverse section.

centrifugal: outwards (from the inside) (cf. centripetal).

centripetal: inwards (from the outside) (cf. centrifugal).

chalaza: region of ovule or seed where nucellus and integuments merge, opposite micropyle.

chlorenchyma: photosynthetic tissue; specialized parenchyma containing chloroplasts.

chlorophyll (chlorophyll a or chlorophyll b): complex magnesium porphyrin compounds forming green photosynthetic pigment contained within chloroplasts.

chloroplast: plastid containing chlorophyll, the site of photosynthesis.

coenocyte: multinucleate cell; i.e. one in which cell division has occurred without cell wall formation (e.g. non-articulated laticifer).

coleoptile (in seedlings): parenchymatous sheath enclosing plumule.

coleorhiza (in monocot seedlings, especially grasses): parenchymatous sheath covering primary root.

collateral vascular bundle: one with xylem and phloem adjacent to each other.

collenchyma: strengthening tissue, consisting of groups of axially elongated, tightly-packed cells with unevenly thickened walls.

colpus (pl. **colpi**): aperture in pollen grain wall, aligned equatorially during the tetrad stage.

companion cell: parenchymatous cell associated with sieve tube element in phloem.

compitum: opening in transmitting tissue of ovary, near micropyle.

cork: suberinized tissue (periderm).

cortex: region in stems and roots between epidermis and central vascular region.

cotyledon: first leaf of the embryo.

cuticle: non-cellular layer of a fatty substance (cutin) that is resistant to enzymes, covering surface of epidermis.

cystolith: calcareous body found in epidermal cell, or in leaf mesophyll.

cytokinesis: cytoplasmic cleavage following nuclear division.

diacytic: one of the classification types of mature stomata; with one or more pairs of subsidiary cells with their common walls at right angles to the guard cells.

diarch root: one with two protoxylem poles.

dictyosome (Golgi body): cell organelle associated with secretory activity.

distal: situated away from the centre of a body or its point of attachment; sometimes terminal on axis (cf. proximal).

domatia: specialized pockets or tufts of hairs on some leaf surfaces, providing shelter for small insects; sometimes associated with extrafloral nectaries.

dorsiventral (bifacial) leaf: one with surfaces morphologically different from each other.

druse: cluster crystal, or compound crystal.

ectomycorrhizal: (fungal mycelium on roots) associated with root
    surface (cf. endomycorrhizal).

ectexine (sexine): outer, sculptured part of exine in pollen
    grain wall.

egg apparatus (in mature embryo sac): egg cell and two
    synergids.

egg cell: part of megagametophyte; haploid cell at micropylar
    end of mature embryo sac that will fuse with sperm to
    form a zygote.

elaiophore: oil-secreting trichome or tissue in flower.

embryo sac: megagametophyte.

embryogenesis: embryo development. Somatic embryogenesis
    is the induction of an embryo-like structure in cell suspension
    cultures and on the surface of callus cultures.

endexine (nexine): inner layer of exine in pollen grain wall.

endocarp: inner layer of fruit wall (pericarp).

endodermis: innermost cell layer of cortex (mainly in roots);
    initially often with Casparian strip, later often thick walled.

endogenous: of deep-seated (internal) origin (cf. exogenous).

endomycorrhizal: (fungal mycelium on roots) invading tissues
    and cells (cf. ectomycorrhizal).

endoplasmic reticulum (ER): continuous membrane-bound
    system of flattened sacs and tubules permeating cell cytoplasm,
    sometimes coated with ribosomal particles.

endosperm: seed storage tissue, formed by fusion of one sperm
    cell with two polar nuclei (i.e. usually triploid).

endotegmic seed coat: one with thickened, mechanical layer
    derived from inner epidermis of inner integument.

endothecium: anther wall layer immediately within epidermis;
    often possessing characteristic thickenings.

epiblast: in grass embryos, outgrowth opposite scutellum.

epicotyl: seedling axis above cotyledons.

epidermis: outermost layer of cells, covering entire primary plant
    surface.

epigeal germination: seedling germination type in which cotyledons are green and borne above ground.

epigynous flower: one with inferior ovary (i.e. the ovary is attached to the receptacle above the level of insertion of the stamens and perianth parts).

epistomatic (leaves): possessing stomata on adaxial surface only.

epithem: tissue (often loosely packed parenchyma) in hydathode between epidermis and vascular tissue.

equatorial (pollen grain aperture): located at or crossing a line midway between the two poles of a microspore or pollen grain.

exarch root: one that matures centripetally.

exine: outer coat of pollen grain, often differentiated into outer ectexine and inner endexine.

exocarp: outermost layer of pericarp.

exodermis: outer few cell layers of root cortex that have become thicker walled and lignified.

exogenous: of superficial (external) origin (cf. endogenous).

exotestal seed coat: one with mechanical layer formed from outer epidermis of outer integument.

fibre (fiber): axially elongated, thick-walled cell, usually occurring as part of a group, lacking contents at maturity, and with simple pits.

fibre-tracheid: cell type that is transitional between a fibre and a tracheid, possessing bordered pits.

filament (in flower): stalk of stamen.

filiform apparatus: wall thickenings in synergid cells of mature embryo sac.

funicle (funiculus): stalk attaching ovule to placenta in ovary.

fusiform: elongated with pointed ends.

generative cell: part of microgametophyte; divides (usually within pollen tube) to form two sperm cells.

girder (bundle sheath extension): in leaves, a group of cells (parenchymatous or sclerenchymatous) linking a vascular bundle sheath with either or both epidermises.

graft: union (by cell differentiation) of tissues of two different individuals so that one (the scion) can survive on the other (the stock).

ground tissue: unspecialized parenchymatous tissue enclosed within epidermis.

growth ring: (in secondary xylem) a distinct growth increment caused by differential rates of growth during a growing season.

guard cell: one of a pair of stomatal cells that surround a pore.

guttation: secretion (usually passive) of water droplets, often at hydathodes.

gynoecium (in flower): collective term for carpels; including ovary, style(s) and stigma(s).

hair (trichome): epidermal appendage.

haustorium (pl. **haustoria**): cellular process that penetrates adjacent tissues and plays a role in nutrient transport. For example, haustoria of parasitic plants are modified roots; endosperm haustoria are specialized absorptive cells.

hemicellulose: carbohydrate constituent of plant cell walls; heterogeneous group of polysaccharides.

heterocellular ray: one composed of cells of different shapes and sizes.

hilum: scar on seed indicating point of attachment of funicle to ovary wall.

histogenesis: tissue differentiation.

homocellular ray: one composed of cells of similar shape and size.

hydathode: region of secretion of water droplets (usually on leaf margin).

hydrophyte: water plant, sometimes displaying specialized features (e.g. aerenchyma).

hypocotyl: seedling axis bearing cotyledons and shoot apex.

hypodermis (often applied to leaves): distinct cell layer(s) immediately within epidermis.

hypogeal germination: type in which cotyledons remain enclosed in seed coat after radicle has emerged.

hypogynous flower: one with superior ovary; i.e. ovary attached to the receptacle above level of insertion of stamens and perianth parts.

hypophysis: uppermost cell of the suspensor.

hypostase: proliferation of nucellus at chalazal end of embryo sac.

hypostomatic (leaves): possessing stomata on abaxial surface only.

idioblast: cell that differs from cells of surrounding tissue.

included phloem: region of phloem embedded in secondary xylem.

inaperturate (microspore or pollen grain): one lacking a clearly defined aperture on the surface, so that the pollen tube can potentially emerge at any point.

integument: either one or two structures (inner and outer integuments) ensheathing the embryo sac, around the nucellus.

intercalary growth (intercalary meristem): diffuse cell division separate from apical meristem or other well-defined meristems.

internode: region between two nodes on stem.

intine: inner layer of pollen grain wall.

isobilateral leaf: one with both surfaces similar, or with palisade tissue on both sides.

Kranz anatomy: (in some plants with C_4 photosynthesis) distinctive leaf anatomy with mesophyll cells radiating from bundle sheaths.

lateral roots: branches of tap root, of endogenous origin.

latex: complex fluid contained within laticifers, consisting of a suspension of fine particles; used to form commercial rubber.

laticifer: latex-secreting cell.

lenticel: region of loose cells in periderm (bark).

lignin: strengthening material deposited with cellulose in plant cell walls, giving rigidity; a high polymer composed of several different types of phenyl-propane units.

ligule: outgrowth of abaxial epidermis in region between sheath and petiole (or between sheath and lamina if petiole absent); derived from cross zone in leaf primordium.

lysigenous (ducts or cavities): developing by cellular breakdown (lysis). (cf. schizogenous).

macrosclereid: elongated sclereid often located in seed coat.

megagametophyte (female gametophyte): mature embryo sac, most commonly consisting of seven cells and eight nuclei (two synergid cells, an egg cell, three antipodal cells, two polar nuclei).

megaspore: female haploid cell resulting from meiosis; usually one of four (or two), of which only one is functional.

megasporogenesis: process of megaspore formation from megaspore mother cell (megasporocyte).

megasporocyte (megaspore mother cell): a diploid cell that will give rise to (usually) four haploid megaspores following meiosis.

meiosis (I and II): two successive divisions of a diploid nucleus to form a haploid gamete.

meristem: region of cell division and tissue differentiation (e.g. apical meristem, intercalary meristem, lateral meristem, vascular cambium, primary and secondary thickening meristems).

meristemoid: isolated meristematic cell, usually the smaller cell resulting from an asymmetric division (e.g. guard cell mother cell).

mesocarp: middle layer of pericarp.

mesogene cell: stomatal subsidiary cell derived from meristemoid.

mesomorphic: displaying no xeromorphic or hydromorphic characteristics.

mesophyll: ground tissue of leaf; mainly consisting of parenchyma or chlorenchyma; often differentiated into palisade and spongy mesophyll.

mesophyte: plant growing in conditions of relatively continuous moisture and/or other nutrients.

metaxylem: primary xylem formed after protoxylem.

microfibril: thread-like component of cell wall, primarily cellulose.

microgametophyte (male gametophyte): cellular component of mature pollen grain, in angiosperms consisting of vegetative cell and generative cell.

micropyle: opening at one end of ovule, usually surrounded by integuments.

microsporangium: pollen sac, contained within anther.

microspore: individual haploid cell that will give rise to microgametophyte. Undergoes unequal mitotic division to form vegetative and generative cells.

microsporocyte: a diploid cell that will give rise to four haploid microspores following meiosis.

microsporogenesis: developmental process leading to production of four haploid microspores from a diploid microsporocyte by meiosis and cytokinesis.

middle lamella: layer between walls of neighbouring cells.

mitochondrion (pl. **mitochondria**): cytoplasmic organelle.

mitosis: cell division to form two cells of equivalent chromosome composition to parent cell; involving four main stages: prophase, metaphase, anaphase and telophase.

mucilage (slime): strongly hydrophilic polymer containing polysaccharides.

multiseriate: consisting of more than one layer or row of cells.

nectary (floral or extrafloral): localized cell or cells that secrete a sugary liquid (nectar).

node: part of stem where leaves are attached.

non-articulated laticifer: one composed of a single multinucleate coenocytic cell.

nucellus: ovule cell layer(s) immediately surrounding megagametophyte.

obturator: proliferation of (usually) ovary tissue near micropyle; secretes substances that guide growing pollen tubes into micropyle.

ontogeny: development; differentiation and growth.

organogenesis: development of organs.

osmophore: scent-producing gland in flower.

osteosclereid: bone-shaped sclereid.

papilla (pl. **papillae**): epidermal appendage; small unicellular trichome.

paracytic: classification type of mature stomata, with one or more subsidiary cells present at either side of the guard cells.

paratracheal parenchyma: axial parenchyma associated with vessels (in secondary xylem).

parenchyma: tissue composed of unspecialized thin-walled cells with living contents.

passage cell: endodermal cell that remains thin walled compared with neighbouring endodermal cells.

pearl gland (pearl body): secretory leaf emergence or trichome that provides food (carbohydrates, lipids and proteins) for ants.

perforated ray cell: (in secondary xylem) ray cell linking two vessel elements, and itself resembling and functioning as a vessel element.

perforation plate: opening in end wall of vessel element; may be simple (a single opening) or scalariform (with bars), or more rarely, reticulate (mesh-like) or foraminate (with pores).

perianth: outer sterile part of flower, consisting of whorls of sepals (calyx) and petals (corolla), or sometimes undifferentiated tepals.

pericarp: fruit wall.

periclinal: parallel with plant surface.

pericycle (in roots): layer of thin-walled cells within the
endodermis.

periderm: cork tissue.

perigene cell: stomatal subsidiary cell derived from cell adjacent to
meristemoid, not from the meristemoid itself.

periplasmodium: coalescent mass in anther locule, formed from
protoplasts of tapetal cells.

perisperm: food-storage tissue in the seed, derived from part of
nucellus.

phellem: external derivatives of phellogen.

phelloderm: internal derivatives of phellogen: cork cambium,
or cork meristem.

phloem: tissue that transports food in the form of assimilates;
either primary (produced by apical meristem) or secondary
(produced by vascular cambium).

photosynthesis: process driven by light energy, in which carbon
dioxide is reduced to carbohydrate form, with concomitant
oxygen release.

phyllotaxis: pattern of arrangement of organs on an axis,
e.g. leaves on a stem, flowers on an inflorescence.

pith: central parenchymatous region of stems, often breaking
down to form a cavity.

pit: thin area of the primary and secondary cell wall, often
corresponding with a pit in adjacent cells (forming a pit-pair).

placenta: region of attachment of ovules on ovary wall.

placentation: arrangement of placentae and locules in ovary
(e.g. axile, basal, free central, parietal).

plasma membrane (plasmalemma): cell membrane (within cell
wall) that encloses protoplast.

plasmodesmata: protoplasmic strands passing through primary
pit fields between adjacent cells, and connecting their
protoplasts.

plastid: cell organelle contained within cytoplasm, often with specialized function (e.g. chloroplast, amyloplast).

polar nucleus: one of a pair of nuclei of the mature megagametophyte, often located in a central position.

pollen grain: microgametophyte enclosed by sporopollenin.

pollen tube: tube that emerges from intine of germinating pollen grain.

polyarch root: one with several (more than four) protoxylem poles.

primordium: organ at an early stage of differentiation.

procambium: primary tissue located near (shoot or root) apex that gives rises to primary vascular tissue.

proembryo: early (globular) stage of embryo development, prior to differentiation of cotyledons and hypocotyl.

promeristem (in root apices): region of greatest mitotic activity.

protoplast: living part of cell, surrounded by a plasma membrane.

protoxylem: first-formed primary xylem.

proximal: situated closer to the centre of a body or its point of attachment (cf. distal).

quiescent centre: region of cells at root apex possessing little or no cell division activity.

radicle: first-formed root of seedling.

raphe: stalk attaching seed to ovary (directly derived from the funicle).

raphide: fine, needle-like crystal of calcium oxalate, one of a group of several raphides formed within a single cell.

ray (in secondary xylem): tissue of radially oriented cells, usually parenchymatous, derived from ray initials.

root cap: protective covering of cells over root apex.

root hair: water-absorbing hair on root epidermis.

sarcotesta: fleshy part of seed coat.

scale (peltate hair): specialized trichome, consisting of a fused disc of cells attached to the epidermis by a stalk.

schizogenous (ducts or cavities): developing from intracellular spaces by separation of cell walls. (cf. lysigenous).

schizo-lysigenous (ducts or cavities): developing by a combination of cell separation and degradation.

sclereid: thick-walled sclerenchymatous cell that is either isodiametric (brachysclereid or stone cell), or variously shaped (e.g. astrosclereid, osteosclereid, macrosclereid).

sclerenchyma: strengthening tissue, consisting of cells with thickened lignified walls, usually lacking contents at maturity. (cf. fibre, sclereid).

scutellum: specialized structure in grass embryos, normally interpreted as a modified cotyledon.

sieve tube element: conducting cell in phloem; possessing sieve areas and sieve plates in walls.

sporopollenin: highly resistant complex polymer that forms outer wall of pollen grain.

squamules: appendages (often glandular) that occur in leaf axils.

stigma: secretory region on gynoecium that is receptive to pollen grains.

stipules: appendages at base of leaf sheath, often paired and sometimes leafy.

stoma (pl. **stomata**): epidermal cell complex that facilitates gaseous exchange, usually present on aerial parts of plant; consisting of two guard cells surrounding a pore.

stone cell: isodiametric sclereid.

storied (in secondary xylem, referring to vessels, rays or axial parenchyma): with stratified structure, occurring in rows.

styloid: elongated prismatic crystal of calcium oxalate.

suberin: lipophilic substance (similar to cutin) deposited in walls of some cell types, e.g. in cork cells.

subsidiary cells: epidermal cells adjacent to stomata that differ from other pavement epidermal cells.

sulcus (pl. sulci): aperture in pollen grain wall, located on its distal face.

suspensor: row of cells attaching globular proembryo to wall of embryo sac.

symplast: connected living protoplasts of adjacent cells.

syncarpous gynoecium: one in which carpels are (at least partially) fused.

syncytium: cytoplasmic region enclosed by a single plasma membrane and bearing several nuclei.

synergid: part of megagametophyte; one of a pair of cells located at micropylar end of mature embryo sac.

tannins: a group of phenol derivatives.

tap root: main central root, formed directly from seedling radicle.

tapetum: layer of nutritive tissue between microsporocytes and wall of anther locule.

tegmen: inner layer of seed coat, formed from inner integument.

testa: seed coat; or in bitegmic seeds, outer layer of seed coat, formed from outer integument.

tetrad: group of four microspores or megaspores, the daughter cells of a single microsporocyte or megasporocyte.

tetrarch (root): one with four protoxylem poles.

tracheid: xylem water-conducting cell, usually possessing bordered pits but lacking perforation plates.

transfer cell: specialized plant cell that facilitates transport of soluble substances across tissue boundaries.

transmitting tissue (stigmatoid tissue): secretory tissue of style through which pollen tubes grow, from stigma to ovule.

triarch root: one with three protoxylem poles.

trichoblast: root epidermal cell that gives rise to a root hair.

trichome (hair): epidermal outgrowth.

tricolpate (microspore or pollen grain): one with three equatorial apertures (colpi).

tunica-corpus: regions of central shoot apical organization.

tylose: (in secondary xylem) outgrowth of wall of axial parenchyma cell into a vessel element through a pit; eventually blocking water passage through vessel.

unifacial leaf: one with both surfaces similar, sometimes derived from a single (usually abaxial) surface.

uninterrupted meristem: region of diffuse cell divisions that is continuous with the apical meristem; producing extension growth of the axis.

uniseriate: consisting of a single layer or row of cells.

vacuole: cavity.

vascular bundle (vascular trace): axial strand of vascular tissue.

vascular cambium: meristem that prcoduces secondary vascular tissue.

vascular tissue: conducting tissue (phloem and xylem).

vegetative cell: one of two cells of microgametophyte.

velamen: outer dermal layer on aerial roots of some tropical epiphytes such as Orchidaceae and Araceae.

venation: arrangement of vascular bundles in leaf (e.g. parallel or reticulate venation).

vessel element: water-conducting cell of xylem, possessing bordered pits on lateral walls and perforation plates on end walls. Groups of axially linked vessel elements form a vessel.

vestured pitting (in secondary xylem): bordered pits surrounded by numerous warty protuberances.

wax: fatty substance often deposited on surface of cuticle (epicuticular wax).

whorl (in flower): band or ring of organs encircling floral axis; sometimes applied to a region of a similar organ type, e.g. stamen whorl.

wood: secondary xylem.

xeromorphic: showing characteristics that are associated with dry or nutrient-poor environments.

xerophyte: plant that grows in a dry (xeric) or nutrient-poor environment.

xylem: complex water-transporting tissue consisting of several different cell types.

zygote: cell formed by fusion of egg and sperm cells; eventually divides to form the proembryo.

References

1. Anderson, W. R. 1980. Cryptic self-fertilization in the Malpighiaceae. *Science* **207:** 892−893.
2. APG. 1998. An ordinal classification for the families of flowering plants. *Annals of the Missouri Botanical Garden* **85:** 531−553.
3. APG. 2003. An update of the Angiosperm Phylogeny Group classification for the orders and families of flowering plants: APG II. *Botanical Journal of the Linnean Society* **141:** 399−436.
4. Arber, A. 1934. *The Gramineae.* Cambridge: Cambridge University Press.
5. Arber, A. 1937. Studies in flower structure III. On the 'corona' and androecium in certain Amaryllidaceae. *Annals of Botany* **1:** 293−304.
6. Baijnath, H. and D. F. Cutler. 1993. A contribution to the anatomy and surface details of leaves of the genus *Bulbine* (Asphodelaceae) in Africa. *South African Journal of Botany* **59:** 109−115.
7. Bailey, I. W. 1953. Evolution of the tracheary tissue of land plants. *American Journal of Botany* **40:** 4−8.
8. Barlow, P. W. 1975. The root cap. In: *The development and form of roots.* J. G. Torrey and D. F. Clarkson (eds). London: Academic Press, pp. 21−54.
9. Barthlott, W. 1981. Epidermal and seed surface characters of plants: systematic applicability and some evolutionary aspects. *Nordic Journal of Botany* **1:** 345−355.
10. Barthlott, W. and E. Wollenweber. 1981. Zur Feinstruktur, Chemie und taxonomischen Signifikanz epicuticularer Wachse und ähnlicher Sekrete. *Tropische und Subtropische Pflanzenwelt* **32:** 1−67.

11.   Baum, S. B., J. G. Dubrovsky and T. L. Rost. 2002. Apical
      organization and maturation of the cortex and vascular cylinder
      in *Arabidopsis thaliana* (Brassicaceae) roots. *American Journal of Botany*
      89: 908−920.

12.   Beaumont, J., D. F. Cutler, T. Reynolds and J. G. Vaughan. 1985.
      The secretory tissue of aloes and their allies. *Israel Journal of Botany*
      34: 265−282.

13.   Beck, C. R. 2005. *An Introduction to Plant Structure and Development.*
      Cambridge: Cambridge University Press.

14.   Behnke, H. D. 1991. Distribution and evolution of forms
      and types of sieve tube plastids in the dicotyledons. *Aliso* **3:**
      167−182.

15.   Boesewinkel, F. D. and F. Bouman. 1984. The seed: structure.
      In: *Embryology of Angiosperms.* B. M. Johri (ed.). Berlin:
      Springer-Verlag, 567−610.

16.   Bouman, F. 1984. The ovule. In: *Embryology of Angiosperms.*
      B. M. Johri (ed.). Berlin: Springer-Verlag, 123−157.

17.   Brett, D. W. and A. P. Sommerard. 1986. Ultrastructural
      development of plastids in the epidermis and starch layer of
      glossy *Ranunculus* petals. *Annals of Botany* **58:** 903−910.

18.   Casson, S. A. and K. Lindsey. 2003. Genes and signalling in root
      development. Tansley Review. *New Phyologist* **158:** 11−38.

19.   Choat, B., S. Jansen, M. Zwieniecki, E. Smets and N. M. Holbrook.
      2004. Changes in pit membrane porosity due to deflection
      and stretching: the role of vestured pits. *Journal of Experimental
      Botany* **55:** 1569−1575.

20.   Clowes, F. A. L. 1961. *Apical meristems.* Oxford: Blackwell.

21.   Clowes, F. A. L. 1994. Origin of the epidermis in root meristems.
      New Phytologist **127:** 335−347.

22.   Cocucci, A. E. and A. M. Anton. 1988. The grass flower:
      suggestions on its origin and evolution. *Flora* **181:** 353−362.

23.   Corner, E. J. H. 1976. *The Seeds of Dicotyledons.* Cambridge:
      Cambridge University Press.

24.   Cutler, D. F., P. J. Rudall, P. E. Gasson and R. M. O. Gale. 1987.
      Root Identification Manual of Trees and Shrubs. London: Chapman
      and Hall.

25. Daumann, E. 1970. Das Blütennektarium der Monocotyledon unter besonderer Berücksichtigung seiner systematischen und phylogenetischen Bedeutung. *Feddes Repertorium* **80:** 463−590.

26. DeMason, D. A. 1983. The primary thickening meristem: definition and function in monocotyledons. *American Journal of Botany* **70:** 955−962.

27. Eames, A. J. and L. H. Macdaniels. 1925. *An Introduction to Plant Anatomy*. London: McGraw-Hill.

28. Endress, P. K. 2001. The flowers in extant basal angiosperms and inferences on ancestral flowers. *International Journal of Plant Sciences* **162:** 1111−1140.

29. Erdtman, G. 1966. *Pollen Morphology and Plant Taxonomy*. New York: Hafner Publishing Company.

30. Esau, K. 1977. *Anatomy of Seed Plants*. Canada: John Wiley and Sons.

31. Fahn, A. 1979. *Secretory Tissues in Plants*. London: Academic Press.

32. Feldman, L. J. 1984. The development and dynamics of the root apical meristem. *American Journal of Botany* **71:** 1308−1314.

33. Foster, A. S. 1956. Plant idioblasts; remarkable examples of cell specialisation. *Protoplasma* **46:** 184−193.

34. Foster, A. S. and E. M. Gifford. 1989. *Comparative Morphology of Vascular Plants*. New York: W. H. Freeman & Co.

35. French, J. C. 1986. Patterns of stamen vasculature in the Araceae. *American Journal of Botany* **73:** 434−449.

36. Frey-Wyssling, A. 1976. *The Plant Cell Wall*. Third edition. Berlin: Gebrüder Borntraeger.

37. Furness, C. A. and P. J. Rudall. 1999. Inaperturate pollen in monocotyledons. *International Journal of Plant Sciences* **160:** 395−414.

38. Furness, C. A. and P. J. Rudall. 2001. The tapetum in basal angiosperms: early diversity. *International Journal of Plant Sciences* **162:** 375−392.

39. Furness, C. A., P. J. Rudall and F. B. Sampson. 2002. Evolution of microsporogenesis in angiosperms. *International Journal of Plant Sciences* **163:** 235−260.

40. Furness, C. A. and P. J. Rudall. 2003. Apertures with lids: distribution and significance of operculate pollen in monocots. *International Journal of Plant Sciences* **164:** 835−854.

41.   Furness, C. A. and P. J. Rudall. 2004. Pollen aperture evolution —
      a crucial factor for eudicot success? *Trends in Plant Science* **9:**
      1360—1385.
42.   Galatis, B. and P. Apostolakos. 2004. The role of the cytoskeleton
      in morphogenesis and function of stomatal complexes. Tansley
      Review. *New Phytologist* **161:** 613—639.
43.   Gifford, E. M. and G. E. Corson. 1971. The shoot apex in seed
      plants. *Botanical Review* **37:** 147—229.
44.   Graven, P., C. G. De Koster, J. J. Boon and F. Bouman. 1996.
      Structure and macromolecular composition of the seed coat of
      the Musaceae. *Annals of Botany* **77:** 105—122.
45.   Grootjen, C. J. and F. Bouman. 1988. Seed structure in Cannaceae:
      taxonomic and ecological implications. *Annals of Botany* **61:**
      363—371.
46.   Hagemann, W. 1973. The organization of shoot development.
      Revista Biologia (Lisbon) **9:** 43—67.
47.   Heslop-Harrison, H. and Y. Heslop-Harrison. 1982. The
      specialised cuticles of the receptive surfaces of angiosperm
      stigmas. In: *The Plant Cuticle*. D. F. Cutler, K. L. Alvin and
      C. E. Price (eds). London: Academic Press, pp. 99—119.
48.   Heslop-Harrison, J. 1976. The adaptive significance of the
      exine. In: *The Evolutionary Significance of the Exine*. I. K. Ferguson and
      I. Muller (eds). London: Academic Press, pp. 27—37.
49.   Heslop-Harrison, J. and K. R. Shivanna. 1977. The receptive
      surface of the angiosperm stigma. *Annals of Botany* **41:**
      1233—1258.
50.   Heslop-Harrison, J. 1987. Pollen germination and pollen-tube
      growth. *International Reviews in Cytology* **107:** 1—78.
51.   Hickey, L. J. 1973. A classification of the architecture of
      dicotyledonous leaves. *American Journal of Botany* **60:** 17—33.
52.   Holroyd, G. H., A. M. Hetherington and J. E. Gray. 2002. A role
      for the cuticular waxes in the environmental control of stomatal
      development. *New Phytologist* **153:** 433—439.
53.   Horner, H. T. and H. J. Arnott. 1966. Histochemical and
      ultrastructural study of pre- and post-germinated Yucca seeds.
      Botanical Gazette **127:** 48—64.

54. Howard, R. A. 1974. The stem—node—leaf continuum of the Dicotyledonae. *Journal of the Arnold Arboretum* **55:** 125—181.

55. Iqbal, M. 1995. Structure and behaviour of vascular cambium and the mechanism and control of cambial growth. In: *The Cambial Derivatives*, M. Iqbal (ed.). Berlin: Gebrüder Borntraeger, pp. 1—67.

56. Janssen, B. J., L. Lund and N. Sinha. 1998. Overexpression of a homeobox gene, LET6, reveals indeterminate features in the tomato compound leaf. *Plant Physiology* **117:** 771—786.

57. Jernstedt, J. A. 1984. Root contraction in hyacinth. I. Effects of IAA on differential cell expansion. *American Journal of Botany* **71:** 1080—1089.

58. Johnson, S. A. and S. McCormick. 2001. Pollen germinates precociously in the anthers of *raring-to-go*, an *Arabidopsis* gametophytic mutant. *Plant Physiology* **126:** 685—695.

59. Johri, B. M. 1992. Haustorial role of pollen tubes. *Annals of Botany* **70:** 471—475.

60. Juniper, B. E. and C. E. Jeffree. 1983. *Plant Surfaces.* London: Edward Arnold.

61. Kaplan, D. R. 1973. The Monocotyledons: their evolution and comparative biology. VII. The problem of leaf morphology and evolution in the Monocotyledons. *Quarterly Review of Biology* **48:** 437—451.

62. Kaplan, D. R. 1975. Comparative developmental evaluation of the morphology of unifacial leaves in the monocotyledons. *Botanische Jahrbücher* **95:** 1—105.

63. Kauff, F., P. J. Rudall and J. G. Conran. 2000. Systematic root anatomy of Asparagales and other monocotyledons. *Plant Systematics and Evolution* **223:** 139—154.

64. Kay, Q. O. N., H. S. Daoud and C. H. Stirton. 1981. Pigment distribution, light reflection and cell structure in petals. *Botanical Journal of the Linnean Society* **83:** 57—84.

65. Knox, R. B. 1984. The pollen grain. In: *Embryology of Angiosperms*, B. M. Johri (ed.). Berlin: Springer-Verlag, pp. 197—271.

66. Kuijt, J. 1969. *The biology of parasitic flowering plants.* Berkeley: University of California Press.

67. Larsen, P. R. 1984. *The role of subsidiary trace bundles in stem and leaf development of the Dicotyledonae.* In: *Contemporary Problems in Plant Anatomy* R. A. White and W. C. Dickison (eds.). London: Academic Press, 109–143.

68. Lersten, N. R. 2004. *Flowering Plant Embryology.* Oxford: Blackwell.

69. Maheshwari, P. 1950. *An Introduction to the Embryology of the Angiosperms.* New York: McGraw-Hill.

70. Mahlberg, P. 1975. Evolution of the laticifer in Euphorbia as interpreted from starch grain morphology. *American Journal of Botany* **62:** 577–583.

71. Mahlberg, P. 1993. Laticifers – an historical perspective. *Botanical Review* **59:** 1–23.

72. McCully, M. E. 1975. The development of lateral roots. In: *The Development and Function of Roots.* J. G. Torrey and D. T. Clarkson (eds.). London: Academic Press, pp. 105–124.

73. Metcalfe, C. R. and L. Chalk. 1983. *Anatomy of the Dicotyledons.* II. Second edition. Oxford: Clarendon Press.

74. Natesh, S. and M. A. Rau. 1984. The embryo. In: *Embryology of Angiosperms.* B. M. Johri (ed.). Berlin: Springer-Verlag, pp. 377–443.

75. Nepi, M. and E. Pacini. 1999. What may be the significance of polysiphony in Lavatera arborea? In: *Anther and Pollen: from Biology to Biotechnology.* C. Clement and J. C. Audran (eds.). Berlin: Springer-Verlag, pp. 13–20.

76. O'Dowd, D. J. 1982. Pearl bodies as ant food: an ecological role for some leaf emergences of tropical plants. *Biotropica* **14:** 40–49.

77. Ormense, S., A. Havelange, G. Bernier and C. van der Schoot. 2002. The shoot apical meritem of *Sinapis alba* L. expands its central symplastic field during the floral transition. *Planta* **215:** 67–78.

78. Parkin, J. 1928. The glossy petal of *Ranunculus.* *Annals of Botany* **42:** 739–755.

79. Parre, E. and A. Geitmann. 2005. More than a leak sealant: the mechanical properties of callose in pollen tubes. *Plant Physiology* **137:** 274–286.

80. Pate, J. S. and B. E. S. Gunning. 1972. Transfer cells. *Annual Reviews in Plant Physiology* **23:** 173–196.

81. Peck, C. J. and N. R. Lersten. 1991. Gynoecial ontogeny and morphology, and pollen tube pathway in black maple, *Acer saccharum ssp. nigrum* (Aceraceae). *American Journal of Botany* **87:** 247−259.

82. Prychid, C. J. and P. J. Rudall. 1999. Calcium oxalate crystals in monocotyledons: structure and systematics. *Annals of Botany* **84:** 725−739.

83. Prychid, C. J., P. J. Rudall and M. Gregory. 2003. Systematics and biology of silica bodies in monocotyledons. *Botanical Review* **69:** 377−440.

84. Punt, W., S. Blackmore, S. Nilsson and A. Le Thomas. 1994. *Glossary of Pollen and Spore Terminology.* Utrecht: LPP Foundation.

85. Rasmussen, H. 1983. Stomatal development in families of Liliales. *Botanische Jahrbücher* **104:** 261−287.

86. Reiser, L. and R. L. Fischer. 1993. The ovule and the embryo sac. *The Plant Cell* **5:** 1291−1301.

87. Rudall, P. J. 1987. Laticifers in Euphorbiaceae − a conspectus. *Botanical Journal of the Linnean Society* **94:** 143−163.

88. Rudall, P. J. 1991. Lateral meristems and stem thickening growth in monocotyledons. *Botanical Review* **57:** 150−163.

89. Rudall, P. J. and L. Clark. 1992. The megagametophyte in Labiatae. In: *Advances in Labiate Science*, R. M. Harley and T. R. Reynolds (eds.). London: Royal Botanic Gardens, Kew, pp. 65−84.

90. Rudall, P. J. 1994. Laticifers in Crotonoideae (Euphorbiaceae): homology and evolution. *Annals of the Missouri Botanical Garden* **81:** 270−282.

91. Rudall, P. J. 1995. *Anatomy of the Monocotyledons.* VIII. Iridaceae. Oxford: Oxford University Press.

92. Rudall, P. J. 1997. The nucellus and chalaza in monocotyledons: structure and systematics. *Botanical Review* **63:** 140−184.

93. Rudall, P. J. 2002. Homologies of inferior ovaries and septal nectaries in monocotyledons. *International Journal of Plant Sciences* **163:** 261−276.

94. Rudall, P. J. and M. Buzgo. 2002. Evolutionary history of the monocot leaf. In: *Developmental Genetics and Plant Evolution*, Q. C. B.

Cronk, R. M. Bateman and J. A. Hawkins (eds.). London: Taylor and Francis, pp. 432—458.

95. Rudall, P. J., W. Stuppy, J. Cunniff, E. A. Kellogg and B. G. Briggs. 2005. Evolution of reproductive structures in grasses (Poaceae) inferred by sister-group comparison with their putative closest living relatives, Ecdeiocoleaceae. *American Journal of Botany* **92:** 1432—1443.

96. Ryding, O. 1992. The distribution and evolution of myxocarpy in Lamiaceae. In: *Advances in Labiate Science*. R. M. Harley and T. R. Reynolds (eds.). London: Royal Botanic Gardens, Kew, pp. 85—96.

97. Sage, R. F. 2004. The evolution of C4 photosynthesis. Tansley Review. *New Phytologist* **161:** 341—370.

98. Sampson, F. B. 2000. Pollen diversity in some modern Magnoliids. *International Journal of Plant Sciences* **161:** S193—S210.

99. Shigo, A. L. 1985. How tree branches are attached to trunks. *Canadian Journal of Botany* **63:** 1391—1401.

100. Skinner, D. J., T. A. Hill and C. S. Gasser. 2004. Regulation of ovule development. *Plant Cell* **16:** S32—45.

101. Steer, M. and J. Steer. 1989. Pollen tube tip growth. *New Phytologist* **111:** 323—358.

102. Steeves, T. A. and I. M. Sussex. 1989. *Patterns in Plant Development*. Cambridge: Cambridge University Press.

103. Stevens, P. F. (2006). Angiosperm phylogeny website. Version **6.** http://www.mobot.org/MOBOT/Research/APweb/

104. Stevenson, D. W. 1980. Radial growth in *Beaucarnea recurvata*. *American Journal of Botany* **67:** 476—489.

105. Stevenson, D. W. and J. B. Fisher. 1980. The developmental relationship between primary and secondary thickening growth in Cordyline (Agavaceae). *Botanical Gazette* **141:** 264—268.

106. Stewart, W. N. and G. W. Rothwell. 1993. *Paleobotany and the Evolution of Plants*, Second edition. Cambridge: Cambridge University Press.

107. Sylvester, A. W., L. Smith and M. Freeling. 1996. Acquisition of identity in the developing leaf. *Annual Reviews of Cell and Developmental Biology* **12:** 257—304.

108. Sylvester, A. W. 2000. Division decisions and the spatial regulation of cytokinesis. *Current Opinion in Plant Biology* **3:** 58−66.

109. Tillich, H. J. 1995. Seedlings and systematics in monocotyledons. In: *Monocotyledons: Systematics and Evolution*, P. J. Rudall, P. J. Cribb, D. F. Cutler, C. J. Humphries (eds.). Royal Botanic Gardens, London/Kew, pp. 303−352.

110. Tilton, V. R. and H. T. Horner. 1980. Stigma, style and obturator of *Ornithogalum caudatum* (Liliaceae) and their function in the reproductive process. *American Journal of Botany* **67:** 1113−1131.

111. Tomlinson, P. B. and M. H. Zimmerman. 1969. Vascular anatomy of monocotyledons with secondary growth − an introduction. *Journal of the Arnold Arboretum* **50:** 159−179.

112. Tomlinson, P. B. 1974. Development of the stomatal complex as a taxonomic character in Monocotyledons. *Taxon* **23:** 109−128.

113. Tucker, S. C. 1972. The role of ontogenetic evidence in floral morphology. In: *Advances in Plant Morphology*, Y. S. Murty, et al. (eds.). Meerut: Sarita Prakashan, pp. 359−369.

114. Twell, D. 2002. The developmental biology of pollen. In: *Plant Reproduction. Annual Plant Reviews* 6. S. D. O'Neill and J. A. Roberts (eds.). Sheffield: Sheffield Academic Press, pp. 86−153.

115. Uhl, N. W. and H. E. Moore. 1977. Centrifugal stamen initiation in phytelephantoid palms. *American Journal of Botany* **64:** 1152−1161.

116. Van Bel, A. J. E., K. Ehlers and M. Knoblauch. 2002. Sieve elements caught in the act. *Trends in Plant Science* **7:** 126−132.

117. Vijayaraghavan, M. R. and K. Prabhakar. 1984. The endosperm. In: *Embryology of Angiosperms*, B. M. Johri (ed.). Berlin: Springer-Verlag, pp. 321−376.

118. Vogel, S. 1974. Ölbumen und Ölsammelnde Bienen. *Tropische und Subtropische Pflanzenwelt* **7:** 285−577.

119. Vogel, S. 1990. *The role of Scent Glands in Pollination: on the Structure and Function of Osmophores*. New Delhi: Amerind.

120. Wang, X. F., Y. B. Tao and Y. T. Lu. 2002. Pollen tubes enter neighbouring ovules by way of receptacle tissue, resulting in increased fruit-set in *Sagittaria potamogetifolia*. *Annals of Botany* **89:** 791−796.

121. Weberling, F. 1989. *Morphology of Flowers and Inflorescences*. Cambridge: Cambridge University Press.

122. Weld, M., J. Ziegler and T. M. Kutchan. 2004. The roles of latex and the vascular bundle in morphine biosynthesis in the opium poppy, *Papaver somniferum*. *Proceedings of the National Academy of Sciences* **101:** 13957—13962.

123. West, M. A. L. and J. J. Harada. 1993. Embryogenesis in higher plants: an overview. *The Plant Cell* **5:** 1361—1369.

124. Wilkinson, H. P. 1979. The plant surface. In: *Anatomy of the Dicotyledons*. C. R. Metcalfe and L. Chalk (eds.). Oxford: Clarendon, pp. 97—165.

125. Willemse, M. T. M. and J. L. Van Went. 1984. The female gametophyte. In: *Embryology of Angiosperms*. B. M. Johri (ed.). Berlin: Springer-Verlag, pp. 159—196.

126. Yadegari, R. and G. N. Drews. 2004. Female Gametophyte Development. *Plant Cell* **16:** S133—141.

127. Yeung, E. C. and D. W. Meinke. 1993. Embryogenesis in angiosperms: development of the suspensor. *The Plant Cell* **5:** 1361—1369.

128. Zimmerman, M. H. and P. B. Tomlinson. 1972. The vascular system of monocotyledonous stems. *Botanical Gazette* **133:** 141—155.

Subject index

Taxonomic index